Meteorology:
An Introduction to the
Wonders of the Weather
Part I

Professor Robert G. Fovell

THE TEACHING COMPANY ®

PUBLISHED BY:

THE TEACHING COMPANY
4840 Westfields Boulevard, Suite 500
Chantilly, Virginia 20151-2299
1-800-TEACH-12
Fax—703-378-3819
www.teach12.com

ISBN 1-59803-626-2

Professor Robert G. Fovell, Ph.D.

Professor of Atmospheric and Oceanic Sciences
University of California, Los Angeles

Robert G. Fovell is Professor of Atmospheric and Oceanic Sciences at the University of California, Los Angeles, where he also serves as cofounder and cochair of the Interdepartmental Program in Mathematics/Atmospheric and Oceanic Sciences. He received his Ph.D. in Atmospheric Sciences from the University of Illinois at Urbana-Champaign.

Among his many duties at UCLA, Professor Fovell serves as the chair of the Faculty Executive Committee of the College of Letters and Science and as faculty undergraduate advisor for Atmospheric and Oceanic Sciences. He is also affiliated with UCLA's Institute of the Environment and the Joint Institute for Regional Earth System Science and Engineering. Professor Fovell teaches courses in atmospheric dynamics and thermodynamics, numerical weather prediction, weather forecasting, and cloud dynamics, among other topics. In 2005, he was awarded a UCLA Distinguished Teaching Award (the Harvey L. Eby Award for the Art of Teaching).

Professor Fovell has published extensively, particularly on the subjects of squall lines and storm dynamics, and has served as an associate editor of the *Monthly Weather Review*. He has also appeared as a commentator for programs on the National Geographic Channel and the Discovery Channel.

Table of Contents
Meteorology:
An Introduction to the Wonders of the Weather
Part I

Meteorology:
An Introduction to the Wonders of the Weather

Scope:

This course answers a number of questions that you may have wondered about since childhood: How high is the sky? Why can't we see at night? How do soda straws work? What causes the seasons? And most important: If you open the emergency exit door of an airplane in mid-flight, will you and everything else on the plane immediately fly out the opening?

The course begins with a theme that, at first glance, seems ridiculous: Nature abhors extremes. We can point to numerous extremes in nature, especially when it comes to the weather: rainfall of 450 inches a year in 1 location in India, typhoons with 250-mph winds, temperatures exceeding 130°F in Death Valley and dropping to −80°F in the Alaskan interior. As we'll learn, though, these extremes represent nature's intolerance of imbalance and her efforts to rectify it.

Lecture One introduces our theme and sketches out our approach to understanding the weather, while Lecture Two builds a foundation for that understanding with the important concepts of temperature, pressure, and density. In Lectures Three and Four, we look at the atmosphere, the ozone layer, and the greenhouse effect. In Lectures Five and Six, we explore how heat moves around and see how the Santa Ana winds develop. Beginning with Lecture Seven and continuing into Lecture Eight, we turn to the topic of atmospheric moisture and find out how air is brought to saturation. In Lectures Nine and Ten, we introduce the air parcel as a useful concept for talking about clouds and learn how thunderstorms form and why they don't form as often as they could. Lectures Eleven and Twelve are devoted to 4 principal forces that determine when, where, and how quickly the winds blow.

In Lecture Thirteen, we look at the 1-cell model and the 3-cell model of global atmospheric circulation. In Lecture Fourteen, we see fronts in action and learn about the life cycle of extratropical cyclones that develop along them, while Lecture Fifteen shows us what's happening higher in the atmosphere that also affects these cyclones. Lecture Sixteen uses many of the concepts discussed to this point in the course to further explore vertical and horizontal wind shear, and

Lecture Seventeen covers the influence of mountains on the atmosphere. With Lectures Eighteen and Nineteen, we learn about radar and follow images of squall lines, thunderstorms, and tornadoes. Lecture Twenty discusses the influence of the ocean on weather and climate and outlines the development and effects of El Niño. We continue with "wild weather" in Lectures Twenty-One and Twenty-Two, which explore tropical cyclones and lightning. Finally, in Lectures Twenty-Three and Twenty-Four, we close the course with a look at forecasting and modeling and answer the important question about opening the airplane exit door in mid-flight.

Lecture One
Nature Abhors Extremes

Scope:

Our theme for this course is: Nature abhors extremes. At first glance, this statement seems ridiculous. Nature is replete with extremes, especially extremes of weather. As we'll see throughout the course, however, extremes occur as nature attempts to correct an imbalance or release stress. This lecture, we'll touch on a few of the concepts we'll examine in greater detail throughout the course and look at what's been called the "perfect storm" to see the relationship between low pressure and wind speed and illustrate what happens along a front.

Outline

I. Our theme for this course is: Nature abhors extremes.

   A. At first glance, this statement seems ridiculous. Nature is replete with extremes, especially when it comes to the weather. Around the world, we find extremes of rainfall, drought, temperature, wind speed, and other factors.

   B. By and large, these extremes are symptoms of an imbalance or a stress, which nature hates.

   C. Consider the Earth/atmosphere/ocean system. The radiation from the Sun warms the ground, and the atmosphere is heated from below, which results in an imbalance: cold air, warm ground. Stress builds as nature tries to distribute the heat, resulting in a storm.

   D. Throughout this course, we look at examples that help us understand specific concepts. A can of electronic duster, for example, introduces us to heat conductivity, the ozone hole, greenhouse gases, heat transfer, adiabatic expansion cooling, diabatic processes, phase changes, latent heat, and other concepts.

II. Now, let's talk about water vapor and shear, 2 important ingredients in the storm recipe.

    A. Shear is wind change over a distance—wind speed—wind direction, or both. We're talking here about horizontal wind, which can have shear in both the vertical and the horizontal direction.

    B. An infrared satellite image shows horizontal wind shear as a series of hooks that grow in a chain. This is an example of a shear instability. Such instability creates spin.

    C. Let's now consider vertical wind shear. If you spill a heavy fluid on your kitchen table, it spreads, owing to its own weight, and the leading edge of the fluid is clearly defined. By spilling the liquid, you've made a front, which is a place where fluids of different density meet.

    D. Let's look at an example with cold and warm air. As we see, the more dense cold air burrows beneath the less dense warm air, creating vertical shear and, in turn, spin.

III. Many of you may know what has been called the "perfect storm" from the movie and book of the same name. This storm caused hundreds of millions of dollars of damage from Florida to Maine in 1991.

    A. Tracking this storm at various times on a topographic map, we note that the sustained winds increase as the central pressure drops, but this relationship is not always perfect. Pressure is important, but pressure gradients (or differences) largely determine wind speed.

    B. The perfect storm was a direct ancestor of another hurricane named Grace. Notice that the air flow around Grace is counterclockwise. The forces involved here are pressure gradient force, which drives the wind; Coriolis force; and the centripetal force. They combine to produce counterclockwise motion around cyclones (pressure lows), parallel to isobars (lines of equal sea-level pressure).

    C. The front we see is a meeting place between cold and warm air. As the storm develops, the cold polar air pushes southward, creating a cold front. To the east, warm air is gaining the upper hand, creating a warm front. Along the front, the spin is increasing, and spin creates low pressure.

D. Further into the storm, the cold front and Grace are on a collision course. Meanwhile, a new cyclone has appeared between the cold and the warm fronts. This is the perfect storm.

E. Thirty hours later, Hurricane Grace is a memory, and the perfect storm is evolving a more complicated structure. Jumping ahead again, the perfect storm is sitting over the Gulf Stream, soaking up moisture and energy. It is now a tropical cyclone, a hurricane just like Grace. As for the front, new cyclones will form along it in other places at other times. This is a cycle that will never end.

IV. Our course will consist of an inquisitive approach to the weather.

A. We'll ask and answer many questions, such as: How do the winds blow? Why do they blow the way they do? How are storms formed? Why do storms exist?

B. Our goal throughout the course will be to enhance our appreciation for nature by increasing our understanding of its design.

Suggested Reading:

There is no suggested reading for this lecture.

Questions to Consider:

1. Would the spray can employed in this lecture have gotten cold upon release of its contents if it only contained compressed air?

2. Keeping my motto "Nature abhors extremes" in mind, identify non-meteorological situations of stress or imbalance that could provoke dramatic and perhaps deadly consequences as they strive to regain some kind of balance or equilibrium.

Lecture One—Transcript
Nature Abhors Extremes

Hello, and welcome to Meteorology, our survey of the wonders of the weather. My name is Robert Fovell. I am a Professor of Atmospheric and Oceanic Sciences at the University of California, Los Angeles, better known as UCLA, and I would like to introduce to you my theme for the course. Three simple words—Nature abhors extremes. Nature abhors extremes. This is my motto, the lens through which I view the physical world around us. I would like you to consider this statement, but don't accept it uncritically. After all, it seems like a ridiculous statement. Nature is replete with extremes. It seems to revel in them, especially when it comes to the weather.

Here are a few of Nature's extremes: rainfall. There's a place in India that averages 450 inches of rain a year. There, a drought probably means when the rainfall accumulation is less than 300 inches. But in contrast, in some parts of the Atacama Desert of Chile, it hasn't rained in centuries. Winds. Close to 200 mph winds were clocked in 1979's Typhoon Tip. A 231 mph wind was recorded in April, 1934, during a winter storm at Mount Washington, New Hampshire. Over 300 mph was detected in a devastating Oklahoma City tornado in 1999. Tornadoes strike the United States about 1500 times per year.

Temperatures. The highest temperature ever recorded in the United States was 134°F in Death Valley in 1913. This was a world record high temperature for 9 years. The lowest temperature in the United States—you've guessed it—in the Alaskan interior was about −80°F. Large temperature swings can occur when fronts pass or when winds shift, particularly when they start blowing downslope. In 1943, the temperature in a town in the Black Hills area of South Dakota rose by 49°F in 2 minutes. But nothing matches the temperature and the temperature swings of air heated by lightning. In lightning, the air temperature immediately adjacent to the lightning bolt increases to about 50,000°F, literally in a flash. The air jumps away from the lightning bolt so fast that it breaks the sound barrier, and that's the thunder that we hear. Lightning strikes the Earth millions of times a day.

I can keep going, but why bother? Nature has many more extremes, and the real question is, how can I have the temerity, the audacity to even suggest that Nature abhors these extremes that it is so full of?

Here's why. By and large, those extremes that I have mentioned were symptoms, symptoms of an underlying disease. Look at the word "disease." Dis-ease. It means an uneasiness with, a frustration with, an impairment, an imbalance, a stress. Nature hates stress. Let me do a non-atmospheric example—earthquakes. I'm not a geologist, but living in Los Angeles has given me some unfortunate empirical experience with earthquakes. As you know, the Earth's crust is fractured, like an eggshell, and the pieces are in motion. Some pieces are trying to climb over other pieces. Others are just trying to pass by like ships in the night. All of this motion is frustrated by friction. So the stress builds, and it builds and builds and increases to an intolerable point, and then the fault breaks. The ground lurches. Landscapes and lives are rearranged. When the dust settles, we need to ask ourselves a question. We must consider whether any good has come from it at all. The answer is yes, somewhere there's less stress, for a while, but it'll just happen again.

Now, let's consider the Earth/atmosphere/ocean system. We're receiving a lot of energy from the Sun in the form of radiation, but much of this radiation, much of this energy, streams through the atmosphere with very little absorption. You can't warm the air if it doesn't absorb the radiation. Instead, the radiation warms the ground, so the atmosphere is heated from below like a pot of water on the stove. There's an imbalance—cold air, warm ground. Nature wants to distribute the heat. First, it tries via a process called "conduction," molecule to molecule to molecule—far too slow. So the stress builds. The air stays cold. The ground gets hotter. The stress builds and builds and builds and then a threshold is passed. And suddenly, there's motion, air rushing up and down at high velocity, up to 100 mph, and rushing around and around at high velocity, 100, 200, 300 mph. Storms are what happens when Nature loses its patience. The stress has built to intolerable levels. Something had to be done.

We're talking about imbalance. With lightning, the stroke occurs when an electric charge imbalance becomes too large. Many other atmospheric phenomena are responding to temperature imbalances, variations with height, which I've just mentioned, also variations between land and sea, and between equator and pole. This is the first of many times in this course I will say this very important concept, probably the most important concept in the entire course. Temperature differences make pressure differences. Pressure differences drive winds. Temperature differences make pressure differences, and pressure differences drive winds.

The purpose of winds is to reduce the temperature differences that made the pressure differences to begin with. The goal of winds is to blow themselves out, but—and you knew there was a "but" coming right?—Nature is frustrated. It's frustrated by the lack of absorption of sunlight by air. It's frustrated by the slowness of conduction. It's frustrated by the spherical shape of the Earth. It's frustrated by Earth's rotation. That's a big one. Each of these things helps create stress. Each of these permits temperature differences to grow and grow and grow until Nature has a tantrum. And what good did it do? Like the earthquake, it has brought temporary relief, and as soon as it's over, it just starts all over again.

Here's my working roadmap for the course. Through a few examples, I'll introduce a lot of concepts that I'll explain through the course. A lot of this may not make a lot of sense at first, and it may not be obvious why what I'm saying is important, but if you return to this lecture after finishing the course, you'll see that it all makes sense. For the first example, I want to use this can of electronic duster. This is the kind of compressed fluid or compressed substance that we use to safely blow out dust from electronic equipment. The can is at room temperature, like everything else in this room, but it feels cold to the touch, in the way that the tabletop or the glass or the cup does not, because it's made of metal. Metal is a good conductor of heat. Heat conductivity and its role in the atmospheric heat transfer will be discussed in Lecture Five.

Now, products like this used to use chlorofluorocarbons, or "CFCs." CFCs have been shown to worsen the ozone hole, which we'll discuss in Lecture Three. Nowadays, this product contains a substance called "hydrofluorocarbon." It's ozone safe, but it's also a greenhouse gas. We'll discuss the greenhouse effect in Lecture Four. Now, normally, hydrofluorocarbons are gases, but the contents of this can are at very high pressure, so inside it's actually a fluid. Now, I'm going to start releasing this product and talk while I'm releasing it. As I'm releasing it, the hydrofluorocarbon rapidly converts to vapor. In fact, it's boiling. In Lecture Two, we'll discuss the concept of pressure, and pressure and boiling we'll discuss in Lecture Eight. The pressure drop, as the fluid is coming out of the container, is allowing expansion. This allows and creates cooling without heat transfer. We'll first encounter adiabatic expansion cooling in Lecture Six.

"Adiabatic" is a Greek word. There are lots of Greek and Latin terms in this course, but I'll explain what each of them means. "Adiabatic" is the Greek word that means impassable. We'll see why that actually makes sense. So as I release the contents, one of the things that I notice is the can gets very, very cold. Now, be careful. It's actually possible that if you discharge it too long, your fingers could actually stick to the can and cause frostbite. This is far, far colder than just adiabatic expansion cooling. This is a diabatic process. Diabatic is the opposite of adiabatic, and in this case, it means it involves significant heat transfer. In Lecture Eight, we'll see that an important source of heating and cooling occurs when substances change phase, such as liquid to gas, as in this example.

The vaporizing hydrofluorocarbon took heat from the can and from the air, and that's why the can is so cold. Where did the heat go? It's now hidden, or latent, in the molecular structure of the gas. We call this "latent heat." Our phase changes and latent heat transfers will involve water substance. Speaking of water, let's look at the can. If you look closely at the can, you'll notice that it's actually wet. It might actually have frost on it, it is so cold. Dew has formed. The room air has been chilled to its dew point on contact with the can. We'll talk about this in Lecture Eight. With dew, latent heat was transferred from the condensing vapor to the room air, so my can actually helped warm up this room a little bit, as if it needed it. Latent heat from condensing vapor is storm fuel. Storm fuel for isolated cumulus clouds we'll discuss in Lecture Nine, to long-squall-line storms and tornadic rotating supercell storms of Lectures Eighteen and Nineteen, to hurricanes, which we'll discuss in Lecture Twenty-One. We got a lot out of this little can, didn't we?

Next, let's talk about water vapor and shear. Water vapor and shear are important ingredients in the storm recipe. Water vapor always helps. Shear is sometimes good, sometimes not. Shear is wind change over a distance, wind speed, wind direction, or both. By wind, I'm talking about the horizontal wind, and the horizontal wind can have shear in the vertical direction and the horizontal direction as well. This particular example is horizontal wind shear. What you're looking at is infrared satellite imagery from the National Oceanic and Atmospheric Administration, NOAA. IR, or infrared, is part of the electromagnetic spectrum. We'll see that in Lecture Four.

Specifically, it's a wavelength of infrared radiation that is absorbed by water vapor. Of course, there are no colors like red and blue in the IR part of the spectrum, so these have been added. These are false colors. What you're seeing is a series of hooks that grow in a chain. These hooks are forming, owing to strong horizontal wind shear. This is an example of a shear instability, and instability is what makes the weather so interesting. The reason is because shear creates spin. Spin is ubiquitous in Nature. The ultimate spin is the spin of our planet. The rotation rate of the Earth is 1000 mph at the equator. The spin of the Earth and the spin relative to the Earth are both profound influences on the winds, on clouds, thunderstorms, hurricanes, and beyond.

That was horizontal wind shear, so now let's consider vertical wind shear. Let's say you spill a heavy fluid on your kitchen table. It spreads, owing to its own weight. The leading edge of the fluid is very clearly defined. We've made a front. Fronts are places where fluids of different density meet. Fronts are ubiquitous in Nature. Molasses and air on your table, air and water at the sea surface, oil and vinegar in your salad dressing—fluids with different densities do not want to mix. We can make air denser by cooling it. We'll see the relationship between temperature, density, and pressure in Lecture Two.

Air and water represent a density difference of about 1000 times, but fluids will resist mixing with density differences as small as 1%, and we can make a density difference of 1% in air with a temperature difference of just a few degrees. So let's do an example with cold and warm air, cold air chilled by evaporation of rainwater, maybe cold air chilled by passing over cold ground, something like this. But in any case, the point is that the more dense cold air is burrowing beneath the less dense warm air. Look at the air flows. Do you see the vertical shear? Do you see the spin?

I'm a modeler. I make simulations of the weather and phenomena such as this. This is a simulation of cold air under-running dense air. This creates shear. Shear creates spin. Look what spin can do. In this case, it's creating billows. We call these "Kelvin-Helmholtz billows" or "KH billows." I'll show you this picture again in Lecture Nine. This was a photo taken by one of my former students. These are Kelvin-Helmholtz billows made visible by condensed water, a testament to the power of shear and spin. One thing we'll see repeatedly is that spin creates low pressure. Spin creates low pressure. It's motion that begets new and different motion, motion that helps air rise, motion that helps some storms intensify and others weaken, motion that can make hurricanes deadly, and tornadoes even deadlier.

So far, the shear we've seen is pretty benign, but that's not always so. My next example is the Perfect Storm. The Perfect Storm was a storm that occurred in 1991, and it caused a lot of damage from Florida to Maine and beyond, hundreds of millions of dollars of losses in Massachusetts alone, and 29-foot waves and massive coastal flooding. You may know the Perfect Storm from the movie and the bestselling book of the same title. The book was about fishermen who perished aboard a ship called the *Andrea Gail* on October 28, 1991, when their fishing boat was swamped by a gigantic wave.

The perfect storm is interesting to us because it was so unusual, in evolution and in motion. Let's take a look at the track of this storm. This is a topographic map of northeastern North America and the North Atlantic. The red shading is terrain height above sea level. The red dots show positions of the Perfect Storm at various times. The storm's position and intensity at 18:00Z on October 28, 1991, are marked. In meteorology, we measure time by London time, the Greenwich Meridian, which runs through the city of London. "18:00Z" means 18:00 hours, or 6:00 pm in London. It was 5 hours earlier in the eastern United States, 1:00 pm eastern standard time. Daylight savings time ended the day before—otherwise it would have been 2:00 pm. The storm's sea level pressure at this time is 1006 millibars.

The millibars are traditional units in meteorology. It was also 29.7 inches of mercury, which might be a more familiar unit to you. Is that low pressure? Not really. We'll see it's actually pretty close to the worldwide average sea level pressure. Millibars these days have been replaced by a new unit called the "hectopascal." The hectopascal was named after a great scientist, but to me, it sounds like a disease, so I'll keep my millibars. The plot also shows that the maximum wind speed at this time was 30 knots. A knot is a nautical mile per hour. Our common statute mile per hour is 15% higher, so it's 35 mph. It's also 56 kilometers per hour, or 15 meters per second in other units.

The storm is heading east and south at this time. Radio contact with the *Andrea Gail* was lost just a few hours later, but there was a lot of life left in this storm. Twenty-4 hours later, the storm started turning to the west. It's stronger now, a large pressure drop to 986 millibars. The winds are 50 knots, 58 mph. These are sustained winds. In the United States, that represents a 1-minute average. Gusts could have been a lot higher. On 12:00Z, October 30, noon in London, 7:00 am in Boston, storm pressure reached its minimum of 972 millibars, 34 millibars less than at the time of its birth. Sustained winds are now 60 knots, about 70 mph. You're probably noticing a relationship between pressure and wind speed, sustained winds increasing as the central pressure drops. Do the lowest pressures always produce the strongest winds? Let's watch to see if that relationship changes over time.

These lows are called "cyclones." Air flows around them in circles. We'll see why in Lectures Eleven and Twelve. This cyclone's westward track is very unusual. We'll see in Lecture Thirteen that the mid-latitude winds tend to blow from west towards east, particularly in the middle to upper tropospheric winds that guide lows like that. Those are topics of Lectures Fourteen and Fifteen. You may be wondering what the troposphere is. I'll tell you about that in Lecture Three. The storm weakened during its westward trek. Its central pressure rose to 996 millibars by 18:00Z, October 31. The winds weakened to 40 knots, but the storm's unusual path has led it over the Gulf Stream, and the energy from the warm water has given the storm a new lease of life. The Gulf Stream is part of Earth's grand ocean current system, and we'll discuss that in Lecture Twenty.

The storm then curved east and then north. By 18:00Z on November 1, it's a hurricane. Central pressure is 980 millibars. Sustained winds are 65 knots. Here's proof that a relationship between central pressure and maximum sustained winds is not perfect. The storm's pressure has been lower in the past, but the winds are strongest at this time. Pressure is important, but pressure gradients largely determine wind speed. A pressure gradient is a pressure difference divided by the distance between the points. More about that in Lectures Eleven and Twelve. Our storm is just barely a Category 1 hurricane on the 5-category Saffir-Simpson scale. This particular hurricane went unnamed, sort of, because now it's named the "Unnamed Hurricane." We'll talk more about hurricanes in Lecture Twenty-One, including how hurricanes get their names.

The Perfect Storm didn't come out of nowhere. Its direct ancestor was another hurricane named "Grace," which was moving across the central Atlantic just the day before. Let's take a look. This is a representation of the surface temperature and wind fields at 1:00Z October 28, 17 hours before the Perfect Storm formed off the coast of Nova Scotia. Here's Hurricane Grace. It had just reached hurricane strength, Category 1, 65 knots, 75 mph. Grace was moving to the northwest, as many hurricanes do at that location and time of year. This is a consequence of the large-scale circulation. We'll talk about that in Lecture Thirteen. Grace is surrounded by isobars. These are lines of equal sea level pressure. Grace's central pressure at this time was about 985 millibars. It was strengthening as it continued moving over warm water.

The air flow around Grace is counterclockwise. We'll see why in Lecture Twelve. The forces involved here are pressure gradient force, which drives the wind, Coriolis force, and the centripetal force, and they combine to produce counterclockwise motion around cyclones, parallel to isobars. Actually, if you look carefully, you'll see that the wind actually crosses the isobars at a shallow angle, towards low and away from high. That's due to the fourth force, friction, and we'll talk about why friction turns the winds later. Now, I'm highlighting a front. The front is a meeting place between cold and warm air. Cold air born in the polar regions of the planet air mass is called "mP" and "cP," maritime polar and continental polar. Warm air born in the tropical regions is called "mT" for maritime tropical. We'll talk about air masses in Lecture Thirteen.

We'll see cold, warm, and stationary fronts in Lecture Fourteen. The name depends on which air mass is doing the pushing. At this time, neither air mass is pushing very hard, but that changes quickly. Fifteen hours later, we see the cold polar air is pushing more southward. These are northerly winds. In meteorology, we name a wind by where it has come from. These winds blowing from north towards south are northerly winds. These winds have created a cold front. To the east, warm air is gaining the upper hand. These are southerly winds, blowing from south towards north, and a warm front has been born. Something is happening along the front. The spin is increasing. It's becoming concentrated. Spin makes low pressure.

Let's move ahead another 9 hours. The cold front and Grace are on a collision course. This is a contest that the hurricane is going to lose.

Meanwhile, a new cyclone has appeared along the front, between the cold and the warm fronts. This is the Perfect Storm. It's now been born and the *Andrea Gail* has already been lost. There are 2 cyclones on this figure. Grace is a tropical cyclone, born near the equator, nurtured by warm water, happiest when the temperature gradients are small. We'll see why in Lecture Sixteen, and later, why tropical cyclones hate vertical wind shear. In contrast, the Perfect Storm was born on a front. It loves shear. It feeds on it. It's an extra-tropical cyclone. We'll see its life cycle in Lecture Fourteen.

Ahead 30 hours, it's 7:00Z on October 30. The Perfect Storm is moving west. Hurricane Grace is a memory. The Perfect Storm is evolving a rather more complicated structure than the textbook example we'll see in Lecture Fourteen. It's always good to remember that textbook examples aren't often found outside of textbooks. Let's jump ahead another day and a half to 22:00Z, October 31. The Perfect Storm has finished its move towards the west. It's now sitting over the Gulf Stream, soaking up that moisture and energy. The front is still there, but our storm is not on the front anymore. In fact, it's not even an extra-tropical cyclone anymore. It's a tropical cyclone, a hurricane just like Grace. As for the front, new extra-tropical cyclones will form along it in other places at other times. This is a cycle that will never end.

We are embarking on an inquisitive approach to the weather. We'll ask and answer many questions, the most important being how and why. How the winds blow, why they blow the way they do, how they're born, how storms form and why they exist, how they're born, move, and why they die, why the most important rain in a storm possibly is that which actually never reaches the ground? Some more questions. Why is the sea breeze cool and the Santa Ana wind hot? How do barometers work? Why is it often bumpy as you fly over mountains, and cloudy as well? How can it get so windy in the lee of mountains? Why does dew form? Why is the sky blue and the setting Sun red?

In this course, I will show you how and tell you why, guided at all times by Einstein's famous dictum, "Make things as simple as possible, but no simpler." Some concepts will be easy. Some will take time to marinade. I employ repetition. Be assured that if something's important, I'll repeat it. If something's important, I'll repeat it. See, I just did it. But what I really mean is I'll be returning to an important concept, a theme, an example repeatedly to remind you, to reinforce. At every turn, I'll try to appeal to your intuition and experience. Many things turn out to be simple applications of what we already know, what I call my "kitchen examples." However, the most interesting phenomena are probably those that defy both our intuition and our experience. An example is why is the sky blue at twilight? I'll bet the answer surprises you. We'll discuss that in Lecture Twenty-Two.

Here are a few questions to keep in mind along the way. In the movie *The Day After Tomorrow*, there's a scene in which bitterly cold air, originating many miles above the ground, is drawn down by a powerful vortex, flash-freezing the city of New York. In one scene, a flag is frozen in half flutter. Is that possible? In the opening scene of the movie *Twister*, a grown man is shown being sucked out of an underground shelter by a tornado. Is that plausible? Airplanes fly at a level where the air is bitterly cold. It draws in this air to maintain the cabin pressure. Do airplanes have to run their heaters continually in order to compensate? You're seated on the same airplane. You're seated in an exit row. You're afraid of falling asleep, your hand accidentally opening the emergency hatch, precipitating a midair disaster, in which everyone and everything is sucked out of the airplane, starting with you. Are your fears reasonable? Can you dig a well so deep that you can't draw water from it? How is El Nino like water sloshing back and forth in a bathtub?

My goal is to help you understand Nature. Whenever we understand something, however, we peel away a little bit of the mystery. The English poet John Keats complained that by explaining how it worked, Sir Isaac Newton had "unweaved the rainbow." I could not disagree more. Our appreciation for Nature is enhanced by understanding her design. Please join me as we unweave the mysteries of the weather.

Lecture Two
Temperature, Pressure, and Density

Scope:

In this lecture, we'll consider 3 important concepts: temperature, pressure, and density. We'll see what temperature really measures, why pressure decreases with height, and why density is often the overlooked crucial factor. We'll also see how they're interrelated with the simple, powerful ideal gas law. Finally, we'll consider the implications of nature's desire to move mass from higher to lower pressure and look at the important concept of hydrostatic balance. As we go through the lecture, you might keep in mind the following questions: Why do cold and warm fronts exist? Can you dig a well so deep you cannot pump water from it? How do soda straws really work? How high is the sky? Why do we still have an atmosphere?

Outline

I. Some tools we'll use frequently in this course are temperature, pressure, and density; let's begin with temperature.

  A. Temperature is the microscopic kinetic energy of atoms and molecules, which vibrate and translate even in solids, so long as the temperature is above absolute zero. At absolute zero (−273°C, −459°F, and 0°Kelvin), all microscopic motion ceases.

  B. Pressure is the force per unit area. To create pressure, we apply force. In the atmosphere, force is largely gravity due to the weight of air. Sea level pressure is 15 pounds per square inch, or 30 inches of mercury, or 1000 millibars.

II. To a large degree, pressure represents the weight of down-lying air; therefore, pressure decreases with height.

  A. As we ascend in the atmosphere, more of the mass of the atmosphere is below us and less is above, so the pressure pushing down on us decreases.

  B. Because we know that pressure is proportional to mass, this means that half the mass of the atmosphere is between 1000 and 500 millibars, and 80% of the mass is between 1000 and 200 millibars.

III. Density is mass divided by volume, usually measured as kilograms per cubic meter.

   A. For gases like air, temperature, pressure, and density are related through the ideal gas law: $p = \rho r t$. Here, p is pressure, measured in pascals; ρ is density; t is temperature in the Kelvin scale; and r is a proportionality constant.

   B. The ideal gas law implies that temperature, pressure, and density are not independent; changing one changes one or both of the others.

   C. If we hold density constant, increasing temperature causes pressure to rise.

   D. If we hold pressure fixed, as temperature rises, density decreases. This means that at the same pressure, warm air is less dense than cold air. This is crucial because fluids with different densities resist mixing.

   E. In the atmosphere, fronts represent the meeting places between air masses that have different densities. They push against each other, in part, because they resist mixing.

IV. What is the pressure measure known as the inch of mercury?

   A. The inventor of the barometer was Evangelista Torricelli in the 1600s, but you make a barometer of sorts every time you drink water through a straw.

   B. If you want to drink out of a straw from a glass of water, you have to create a vacuum. Once the vacuum is created, the atmosphere does the job of pushing water up into the straw from below.

   C. The difference between the top and the base of the fluid column indicates how much force is pressing down on the outside of the straw. Force per unit area is pressure.

   D. At standard sea level pressure of 1000 millibars, the atmosphere can support a water column 33 feet high. But rather than use a water column, Torricelli employed mercury for his barometer. Sea level pressure can support a mercury column 30 inches high, hence the unit of pressure.

V. Here's a key concept: Nature seeks to move mass from high to low pressure.

    A. Note that pressure differences can exist in all directions, as you can see in the familiar phenomenon of trapping fluid inside a straw with your finger.

    B. Nature seeks to move mass from high to low pressure. This means nature wants the water to rise in the straw, but it doesn't. Why not?

    C. The missing piece is that water itself has weight, and this weight is pushing downward, due to gravity.

    D. Here's another key concept: If a fluid is not accelerating, then the forces must be balanced. The primary balance here is the pressure difference acting upward and the gravity force pulling down. This is a stalemate we call "hydrostatic balance."

    E. The straw is sealed, and we have hydrostatic balance, 2 powerful opposing forces, but no net motion. When you remove your finger, the forces come into balance, and the water flows out.

Suggested Reading:

Turner, *Scientific Instruments 1500–1900*.

Questions to Consider:

1. We have described pressure as largely being the weight of the overlying air. Actually, anything above us should increase the downward force. But, if that's true, why aren't we discomfited, or even crushed, when a large, very heavy airplane flies overhead? Hint: It has nothing to do with lift.

2. On very hot days, jumbo jets are not permitted to take off or land at some airports, owing to insufficiently long runways. Why?

Lecture Two—Transcript
Temperature, Pressure, and Density

Welcome back to Meteorology. We are unweaving the wonders of the weather. In this lecture, we will see 3 important concepts: temperature, pressure, and density. We'll see what temperature really measures, why pressure decreases with height, and why density is often the overlooked crucial factor. We'll see how they're interrelated with the simple, powerful ideal gas law. We'll see how to make a barometer and situations where you would rather not have made one. We'll see implications of Nature's desire to move mass from higher to lower pressure, and the important concept of hydrostatic balance.

Some opening questions to think about while I'm talking in this lecture: why do cold and warm fronts even exist? Can you dig a well so deep you cannot pump water from it? How do soda straws really work? How high is the sky? Why do we still have an atmosphere? I'll start off mainly using familiar old English units, such as inches, miles per hour and Fahrenheit, but I'll provide modern metric equivalents as well. In my classes at UCLA, I often use old-fashioned units, such as Fahrenheit and yards, and a colleague once criticized me for that. He asked me why I wasn't using the language of science and forcing my students to speak my language. But I want you to understand what I'm talking about. I'm willing to speak your language and in return, you learn mine, if you don't know it already. As time goes on, you'll notice that I'll emphasize metric measures more and more.

The tools we're going to use very frequently in this course are temperature, pressure, and density, so let's look at temperature. Temperature is one of those concepts we may understand better before looking at it too closely. Temperature is the microscopic kinetic energy of atoms and molecules, which vibrate and translate even in solids, so long as the temperature is above absolute zero. At absolute zero, all microscopic motion ceases, and absolute zero is $-273°C$, $-459°F$, and zero on the Kelvin absolute temperature scale. Most of the world, as you know, uses the Celsius or centigrade scale, which dates from the mid-1700s. Like many scales, including the meter and the second, the Celsius scale has been redefined over the years. The 1744 version utilized 2 fixed points, $0°C$, the freezing point of impure water, and $100°C$, water's boiling point, both of those at sea level. Actually, Celsius' original version was inverted. He used 100°C for the freezing point, but that proved unpopular.

In the United States, we persist in using Daniel Gabriel Fahrenheit's temperature scale, invented in 1724. In Fahrenheit's day, the challenge was to create instruments capable of making repeatable measurements, calibrated using common materials. So as his zero point, Fahrenheit used the temperature obtained by a mixture of ice, water, and ammonium chloride, which, no matter what temperatures they start with, always reach the same temperature. He called this "0." Some sources say that he then chose 32°F to be freezing, and 96°F to be human body temperature. That may seem very weird at first, but why did he use 32-degree intervals? The reason is because he made his scale through bisection, and using 32 and 64, which are powers of 2, allowed him to make his scales accurately. Later on, the scale was adjusted to force 180°F to exist between freezing and boiling points, and this shifted human body temperature upwards slightly to the more familiar 98.6°F.

Fahrenheit lacks Celsius' clean and simple rationale, but you know what? It's curiously well suited to meteorology, if only accidentally. The temperatures of 0°F and 100°F bracket the range of temperatures commonly encountered in temperate climates. As an example, at this time, Seattle, Washington's, all-time record high is 100°F, and its all-time record low is 0°F respectively. Of course, we need to keep in mind that records are made to be broken. As a result of their rarity, 100°F and 0°F have become psychologically significant numbers, as we interpret weather extremes on the Fahrenheit scale. A temperature below 0°F isn't just cold—it's subzero cold. A temperature exceeding 99°F isn't just hot—it's triple-digit heat. Somehow, 38°C doesn't seem quite so exceptional.

In this course, I'll try to provide temperatures in both Celsius and Fahrenheit, and big numbers we'll manipulate in Kelvin as well. I admit that I still think in Fahrenheit, so here are some benchmark temperatures that I use for quick conversions. A warm day is 30°C. That's 86°F. A typical indoor temperature might be 20°C. That's 68°F. Note the transposition there: 30°C, 86°F; 20°C, 68°F. And 10°C is 50°F. Freezing is 0°C and 32°F. Minus 40 is the same on both scales. That's a special number, as we'll soon see. Converting between Celsius and Kelvin is more straightforward—just add 273 to Celsius. OK, it's 273.15.

Now, I mentioned that −40 degrees is a special temperature, and it's not because the Fahrenheit and Celsius scales cross at that point. Previously, I emphasized that 0°C, or 32°F, was the temperature at which impure water freezes. Impurities can provide surfaces on which the freezing process can start. Pure water can remain liquid to a much lower temperature, very close to −40 on either scale, and this is called "supercooling." We will see the meteorological implications of supercool liquid water later. The first example is involved in aircraft icing. The Celsius scale is pegged to the boiling point of water, 100°C, but that's the boiling point at sea level. As you ascend above sea level, boiling point decreases, and eventually we'll see why, but the reason involves our next tool, which is pressure.

Pressure is the force per unit area. I'm going to take my hand, and I want to exert a pressure on my hand. Well, I'm going to take my other hand and push it down, and I'm going to start pressing. I'm applying a force over the area of my hand. I have created a pressure. Suppose I want to increase the pressure pushing down my hand at the bottom. How can I do that? There are 2 ways. I can either push down harder, applying a greater force, or I can take the same force or an even smaller force and concentrate it in a smaller area, such as the area of my finger. Now, I'm pressing down with a force in a much smaller area, and I'm certainly feeling the pressure. In the atmosphere, force is largely gravity, due to air's weight. We often say light as air, but you know what? Air isn't light.

Sea level pressure is 15 pounds per square inch. It's applying 15 pounds for every square inch of the Earth's surface. That's about half of the pressure you probably have in your automobile tire. Does that sound like a lot? Well, if so, you're used to it, and if it seems insignificant, if I applied an extra 15 psi on your shoulder, I think you would notice. Gravity force is proportional to mass, so if we double the amount of mass in the atmosphere, the surface pressure would also double. Our sister planet Venus has more than 90 times more atmosphere than we have on the Earth, so we would find surface pressure in Venus a bone-crushing experience.

Average sea level pressure is also 30 inches of mercury. The inch of mercury is a measure with a rich history that we'll discover soon. It's 100,000 pascals. That's the official metric system unit, the pascal. It's 1000 hectopascals, which is that unit that sounds like a disease, and it's 1000 millibars, the traditional unit used in meteorology and the unit I'll use through this course. To a very large degree, pressure represents the weight of down-lying air, as we said, and it's pressing downward due to gravity. Therefore, pressure decreases with height. As we ascend in the atmosphere, more of the mass of the atmosphere is below us and less is above, so the pressure pushing down on us decreases. This means for a surface pressure of about 1000 millibars, somewhere above our heads the pressure is only 750 millibars, 500 millibars, above that, 250 millibars, and ultimately, zero millibars at the top of the atmosphere.

Additionally, since we know that pressure is proportional to mass, this means that half of the mass of the atmosphere resides between 1000 and 500-millibar levels, and 80% of the mass resides between 1000 and 200 millibars. Here's a question. How high is the sky? Let me ask it more prosaically. How thick is the atmosphere? Well, it's rather difficult to precisely identify the top of the atmosphere. Like the proverbial old soldier, the atmosphere just fades away. Suppose we use the 10-millibar level as an example. Then 99% of the mass of the atmosphere is beneath the 10-millibar level, and more than 99% of the weather. It may surprise you that the average level of the 10-millibar surface above our heads is 30 kilometers, or 18 miles, above sea level. Consider that as a horizontal distance. That's about half of the commute between my office and my home. Relative to the Earth's radius of 6400 kilometers, or 4000 miles, the 10-millibar level is less above our head than the thickness of the paper on this globe. The protective blanket that the atmosphere represents is incredibly thin.

Furthermore, air is very compressible. It's easy to squeeze, so it's scrunched at the bottom near the surface, owing to its own weight. We may have 99% of the mass of the atmosphere below 18 miles, but 50% is in the lowest 3.5 miles, or 5.5 kilometers. The average height of the 500-milibar level is only 3.5 miles above sea level. Pressure doesn't just decrease with height—it decreases exponentially with height. And because of air's squeezability, something else also decreases with height: density, which is our next topic.

Density is mass divided by volume. In the metric system, the mass is kilograms and the volume is cubic meters, so it's kilograms per cubic meter. Consider a box containing air or some other material. How would I change the density of stuff in that box? Well, I can cram more stuff into the box, or I can make the box with the same mass smaller. Either will increase the density. For gases like air, temperature, pressure, and density are related through a simple, powerful equation, the ideal gas law: It's $p = \rho\,r\,t$.

So p is pressure, measured in pascals; ρ, a Greek letter, is used for density. Kilograms per cubic meter are the units. And t is temperature in the Kelvin scale; r is just a proportionality constant, unique to each gas or combination of gases. How do we find this value? We can look it up in a book. The important point is the ideal gas law implies that temperature, pressure, and density are not independent. They're not independent. Changing one changes one or both of the others. Let me do 2 examples of the ideal gas law, one that's already intuitive to us and one that is crucially important for the atmosphere.

The first one is if we hold density constant, increasing temperature causes the pressure to rise. Let me demonstrate that with a soup can. Here's a soup can and inside is tomorrow night's dinner, minestrone. I know that while soup is good food, it's not an ideal gas, so the first thing I got to do is open this up, take all the soup out, replace it with air, and seal it back up again. This is a sealed and rigid container. Its mass and volume are fixed and therefore the density is constant. If I start heating this up, what's going to happen to the pressure? You know what's going to happen to the pressure. This can is going to explode if I heat it up too much.

The second example is this. If we hold pressure fixed, as temperature rises, density goes down. This means that at the same pressure, warm air is less dense than cold air. Let me repeat that. At the same pressure, warm air is less dense than cold air, so that means that one way of making air less dense is to heat it up. We can create density differences with temperature differences. This is crucial because fluids with different densities resist mixing. Air and water have a density difference of a factor of 1000. Oil and vinegar in your Italian salad dressing have a more subtle difference, but you know from empirical experience that you can shake them up and they won't stay mixed. In the atmosphere, fronts represent the meeting places between different air masses that have different densities. They push against each other, in part because they resist mixing. We often use terms like "cold front" and "warm front," but the real difference here is density, and air masses will resist mixing even with density differences as small as 1%.

What is this strange pressure measure, the inch of mercury? Its story is the story of the barometer itself. If you look this up in an encyclopedia, it will tell you the inventor of the barometer was Evangelista Torricelli in the 1600s. But I think the name of the inventor of the first actual barometer has been lost to antiquity. He was the first poor soul who had a well so deep that he couldn't pump water from it. You make a barometer of sorts every time you drink water through a straw. Let's see how it works. Picture a water well or a glass partially full of water. The atmospheric pressure is being exerted on the plain surface of that water. It's pushing down everywhere. Now insert a tube or a straw, and notice the water does rise up a little bit, but that's due to surface tension. The water doesn't keep rising farther into the straw because there's also air in the tube, pushing down. We want to get water out of that tube, so the first thing we have to do is get the air out by creating a vacuum.

Once the vacuum is created, the atmosphere does the job of pushing water up into the tube from below. The atmosphere is doing the job of pushing the water up. Now, the difference between the top and the base of the fluid column indicates how much force exists, pressing down on the outside of the straw. Force per unit area is pressure. We have created a barometer. At standard sea level pressure of 1000 millibars, the atmosphere can support a water column 33 feet, or 10 meters, high. If your well water is down deeper than that, suction alone cannot extract the water, and the reason is it isn't suction that was pulling the water up out of the well anyway—it was atmospheric pressure pushing down on the fluid surrounding the tube.

This is counterintuitive. If we apply suction to a tube and pull up water, it is logical to assume that we did it. But when examining a phenomenon in the natural world, we should first identify all the forces acting and then determine the most important force. In this case, I claim the most important force is atmospheric pressure pushing down outside the tube. How can we test this? We can test this by sealing the well or cup around the tube. Now atmospheric pressure cannot be exerted on the fluid around the tube because it's isolated from the atmosphere. Now when we place our lips around the tube, the system is sealed. Try to draw on the tube, but don't try too hard. You can hurt yourself. You can't draw fluid up into that straw far enough to taste it. Now, you may be wondering about juice boxes. They're sealed, or nearly so, but that's not really a fair test because juice boxes are flexible. Even if you don't squeeze on the box, atmospheric pressure is sufficient to do the job. Atmospheric pressure is the key.

I already noted that the standard atmosphere water column could stand 33 feet high. That makes for a very accurate and precise instrument, but it's certainly a very inconvenient one. You can't exactly put it in your back pocket. So Torricelli employed mercury, a liquid metal. Sea level pressure can support a mercury column about 30 inches high, hence the unit of pressure that persists to this day. So far, we've had examples of fluid barometers, but most barometers these days are of the aneroid variety. That's a Greek word, meaning without water or without fluid. Some of these use elastic membranes that respond to pressure variations. You're walking around with 2 of them—your eardrums.

Here is a key concept. The barometer examples demonstrated the importance of pressure in generating motion in fluids such as water and air, but in each case, it wasn't the pressure itself. It was a pressure difference that was truly important. The key concept is this: Nature wants to move mass from high to low pressure. Nature wants to move mass from high to low pressure, from surplus to deficit. Nature responds to pressure differences by trying to eliminate them. She isn't always successful, but Nature is always trying. Further, we need to note that pressure differences can exist in all directions.

Another familiar example illustrates how and why. I have a cup of a colored liquid, and in it, I have a straw. I'm going to do something that will be immediately very familiar to you. I'm going to reach in here and I am going to cover up the top of the straw with my finger, and then I'm going to lift the straw out of the fluid. We see that the fluid is trapped inside the straw. It remains trapped until my finger is released and the water returns to the glass. Why did the fluid remain trapped before I removed my finger? Why did it flow out of the straw once I moved my finger away? Let's take a closer look.

The straw is sealed at the top by my finger, and the straw is open at the bottom, so atmospheric pressure is pressing down on the top. That has no effect because the straw is sealed, sealed by my finger. Pressure is also acting on the sides of the straw as well, which also has no effect because the straw is sufficiently rigid and inflexible. But atmospheric pressure also presses upward from below, and inside the straw, between my finger and the top of the fluid, there's also a small amount of air inside, trapped above the water. It also has weight, and it's pressing down. Although we may think of atmospheric pressure as always pushing down, the straw helps us see the pressure differences upward, so the net pressure force is acting upward. Nature wants to move mass from high to low pressure. This means Nature wants the water to rise in the straw, but the water's not rising. Why isn't the water rising in the straw?

We're not done identifying all the relevant forces. The biggest missing piece is that water itself has weight, and this weight is pushing downward, due to gravity. Here's another key concept. If a fluid is not accelerating—in this case, up or down—then the forces must be balanced. The primary balance here is the pressure difference acting upward and the gravity force pulling down. This is a stalemate we call "hydrostatic balance." "Hydro" is a Greek prefix for water. It means fluid, and "static" means stationary. Hydrostatic balance is the answer to the question, why do we still have an atmosphere? The fact that hydrostatic balance can be broken is the answer to the question, why can we have thunderstorms?

Let's break the stalemate. The straw is sealed and we have hydrostatic balance, 2 powerful opposing forces, but no net motion. But now I've removed my finger and the forces are now in balance, the water flows out. Why? Before I removed my finger, I had a large vertical pressure difference. The entire atmosphere was pushing up from below, and almost nothing was pushing back. But after I removed my finger, the pressure difference in the vertical became very small. The atmosphere was pushing up, but it was also pushing down with almost as much force, and therefore, effectively, cancelling itself out. There was nothing left to stalemate the gravity force, and so the fluid ran out of the straw.

Why do we still have an atmosphere? Another question is why isn't it even shallower than it is? Gravity is a powerful force, relentless, trying to bring the atmosphere to your knees, but the vertical pressure difference is also extremely large. We saw a 500-millibar pressure drop in just 3.5 miles between the surface and midtroposphere, and that makes a force so strong that only gravity can restrain it, and that is hydrostatic balance. It is an epic struggle, a source of great strain. The atmosphere is tightly strung and ready to be plucked. I usually demonstrate hydrostatic balance with a rubber band. I can stretch up hard at the top, and that represents the pressure force that is trying to push the atmosphere off the planet and into outer space. My other hand is pulling down, and that represents gravity, which wants everything to go to ground, and the shallower at the ground, the better. This is a terrific, terrific strain. I'm straining upwards with the pressure force and I'm straining downwards with the gravity force. It is creating a tremendous tension.

What happens if I disturb this balance? What happens if I let go at the top or the bottom of this rubber band? You know what's going to happen: motion, in the vertical, up or down, at tremendous speeds. The hydrostatic balance represents not a gentle situation, where air molecules are just sort of floating around in the sky, but is actually represents a situation of tremendous stress, tremendous strain. The atmosphere is under strain at every single moment and it's always ready to break and it's ready to be plucked.

I mentioned in Lecture One that we'll encounter a number of Latin and especially Greek terms in this course. In Lecture One, we encountered the word "adiabatic," which meant impassible, a term we'll learn much more about in Lecture Six. In this lecture, we've already seen the word "aneroid," which means without fluid in Greek, and "hydrostatic," with the prefix being the Greek word for water. Perhaps I've been remiss in not introducing you to the most relevant Greek term of all, "meteorology." When I tell someone I'm a meteorologist, there are 2 common and swift questions that come in reaction. "Which channel are you on?" I get asked. I'm only on your TV. "Why do you study meteors? After all, isn't meteorology the study of meteors?" Actually, "meteorology" is Greek for the study of things high in the sky, which is why falling rocks from space are called "meteors," even though meteorologists don't study them.

Meteorology is also the title of a book written by Aristotle around 350 B.C., a great advance in its day. Sadly, from the time of Aristotle up until the late 19th century or so, very little progress in our understanding of the weather was made. Only at the dawn of the 20th century did meteorology start becoming a serious science, like physics, like chemistry, and like astronomy. Rapid advancements took place after World War II, and because of World War II as well. For one thing, radar was invented, and then rockets were created, and they launched satellites into outer space, radar and satellites becoming our eyes on the weather. For another, the military realized we had better start getting a clue about weather and storms. And that brings us to today, where most meteorologists call themselves "atmospheric scientists" because we don't like being asked which channel we're on.

Let's summarize. Temperature, pressure, and density are our tools for understanding our atmosphere and its weather. Pressure exists because the air has weight, and pressure decreases with height. Since air is compressible, pressure decreases exponentially with height, as does density. We recognize that fluids with different densities resist mixing—oil and vinegar in your Italian salad dressing, cold air and warm air at fronts. At sea level, pressure is strong enough to stand a column of water on end 33 feet high, but we saw that its pressure differences inside and outside of that column, inside and outside of your soda straw, between the surface and the sky that drives Nature into action. Nature wants to move mass from high to low pressure. If Nature always got her way, we would not even have an atmosphere. It would have been lost to space, and those of us who breathe would miss it. But our atmosphere is the scene of a great struggle between upward pressure differences and the downward gravity force, representing hydrostatic balance. Even on a calm day, the atmosphere is a place that is straining against itself, always this close to snapping, except when it does.

Let's look ahead. We have seen the very important role that pressure differences play in atmospheric motions. A further step is to recognize that while pressure differences drive winds, pressure gradients determine the velocity and the ferocity of the winds. We'll learn much more about pressure gradients in future lectures. First, we need to appreciate what the atmosphere is made of, how it came to be, and why our planet is even habitable at all. We'll start on this in our next lecture.

Lecture Three
Atmosphere—Composition and Origin

Scope:

In this lecture, we'll see that the variation of temperature with height is complex. We'll learn what air is made of, now and in the distant past, and we'll be introduced to the greenhouse effect and its major players. Think about these questions as we go through the lecture: Is it always true that warm air rises and cold air sinks? Could something that comprises only 4 out of every 10,000 molecules in air really make a difference? What good is the ozone layer, and how did a chemical used in common spray cans damage it?

Outline

I. The atmosphere has 4 layers, distinguished by how temperature varies with height.

    A. The lowest layer is called the troposphere; this is our weather sphere. It extends from sea level up to about 7.5 miles. Surface pressure is 1000 millibars, and the pressure at the top of the troposphere (the tropopause) is about 200 millibars. Notice that temperature decreases very quickly with height—from 60°F at the bottom to −80°F at the top.

    B. The next layer is the stratosphere. This is a layer of great stability that impedes vertical motion. The stratopause is about 30 miles above sea level, with pressure of about 1 millibar. In this layer, temperature increases with height instead of decreasing.

    C. Next is the mesosphere. Here, temperature resumes its decrease with height. The mesopause is 55 miles above sea level, and the pressure there is 0.01 millibars.

    D. Finally, the thermosphere extends from 55 miles up to where the atmosphere just fades away. Temperatures there can be thousands of degrees, yet the thermosphere is also a cold place. There's virtually no mass in the thermosphere, but it's a deep layer.

II. We can divide air into 2 categories, dry air and water vapor. Dry air is largely fixed in quantity and well mixed through the atmosphere. Water vapor, in contrast, is extremely variable, in horizontal space, in vertical depth, and in time.

   A. Dry air is 78% nitrogen in the form N_2, 21% oxygen in the form O_2, and 1% argon. Other minor constituents of the dry atmospheric mass include carbon dioxide, CO_2.

   B. Carbon dioxide represents 0.0387% of the dry atmospheric mass, but it has an outsized importance as a greenhouse gas and plays a major role in regulating the temperature of the Earth's surface.

   C. Another minor constituent in the dry atmospheric mass is methane, CH_4, which is about 2 parts per million. Methane is roughly 25 times more potent per unit weight than carbon dioxide as a greenhouse gas and is also increasing over time.

   D. Other minor constituents include nitrous oxide, N_2O, and ozone, O_3. The majority of ozone is in the stratosphere, where local concentrations are 250 times larger than they are in the atmosphere as a whole.

III. Let's look at the natural stratospheric ozone cycle.

   A. The act of absorbing ultraviolet radiation causes an oxygen molecule to split into its individual atoms, which will quickly find a nearby O_2 molecule to form O_3, ozone.

   B. Next, ozone absorbs ultraviolet radiation, which splits it apart. In this process, ozone is both created and destroyed, but there's no net loss.

   C. The ozone hole is an area of depleted stratospheric ozone residing over the South Pole. Excessive ozone hole expansion is caused mainly by chlorofluorocarbons, or CFCs.

   D. CFCs play a dual role as greenhouse molecules and destroyers of stratospheric ozone. Because CFCs are chemically inert, they have very long lifetimes, which allows them to spread globally and loft far upward, reaching the stratosphere.

   E. Ultraviolet radiation from the Sun can break CFC molecules apart, liberating the chlorine atoms.

F. Satellite images show considerable inter-annual variability in the ozone hole, but the trend toward larger and deeper ozone holes is clear.

IV. Let's turn to water vapor.

    A. Water vapor represents 0% to 4% of the total atmospheric mass but is extremely variable in space and time. Water vapor is concentrated near the Earth's surface and the lower troposphere.

    B. The ability of air to hold water vapor is a strong function of temperature. Warm air can hold much more water vapor than cold air. As we have seen, temperature decreases rapidly with height in the lower atmosphere. This is one reason why most of the water vapor is located near the ground.

V. Earth didn't always have the atmosphere we have today.

    A. The early Earth probably had an atmosphere of hydrogen and helium, but Earth's gravity is not strong enough to retain those elements.

    B. Later, Earth developed an atmosphere that had significantly more CO_2 than our present atmosphere and virtually no oxygen.

    C. In our atmosphere, the first free oxygen probably resulted from photodissociation of water vapor by intense solar radiation. Significant oxygen concentrations had to await the evolution of photosynthesis.

Suggested Reading:

There is no suggested reading for this lecture.

Questions to Consider:

1. Why does the ozone hole disappear in summertime?

2. The ionosphere is not an atmospheric layer, but rather a region that extends from the upper mesosphere up to 400 kilometers or so. The structure of this region varies dramatically from day to night. Strong solar radiation ionizes atoms, creating layers that absorb AM radio signals but disappear at night. At what time of day do you expect to be most successful at receiving a distant AM station on your radio?

Lecture Three—Transcript
Atmosphere—Composition and Origin

Welcome back to Meteorology. We're exploring the wonders of the weather. In the last lecture, we realized that temperature measures the microscopic activity of atoms and molecules. Pressure is largely due to atmospheric mass weighing down on us, owing to gravity. Density plays a subtle but important role in the atmosphere because substances with different densities resist mixing. For air, these 3 properties are related through a simple equation called the "ideal gas law." Pressure decreases with height because as you ascend, less mass remains to push down on you, and density decreases with height as well because air is compressible and it squeezes down on itself.

In this lecture, we'll see that the variation of temperature with height is complex. We'll learn what air is made of, now and in the distant past as well, and we'll be introduced to the greenhouse effect and its major players. Some things to think about along the way: We have all heard the saying warm air rises and cold air sinks. Is that always true? Could something that comprises only 4 out of every 10,000 molecules in air really make a difference? What good is the ozone layer? How did a chemical used in common spray cans damage it?

First, let's introduce ourselves to the standard atmosphere. The standard atmosphere is created by averaging from equator to pole, from winter to summer, from day to night, from land to sea. The standard atmosphere has 4 layers, distinguished by how temperature varies with height. In the lowest layer, nearest the ground, we call this the "troposphere." That's from the Greek turning sphere, turning implying change, implying the weather. It's our weather sphere. The troposphere extends from sea level up about 12 kilometers or 7.5 miles or so. We remember that surface pressure is 1000 millibars, and the pressure at the top of the troposphere, which we call the "tropopause," is about 200 millibars. With a surface pressure of 1000 and a tropopause pressure of 200, we remember that this represents 80% of the Earth's atmospheric mass because pressure is proportional to mass.

Further, notice temperature decreases quickly with height, and I mean very quickly—from 60°F or 16°C, at bottom, on average, to −80°F, or −62°C, at the top. That's a 140°F temperature drop over a very short distance. Why? Because we'll see later that the atmosphere, the troposphere, is heated from below. Next up is the stratosphere, from the Latin to spread horizontally. You see words like "stratos" in there. This is a layer of great stability, which impedes vertical motions. The top of the stratosphere is called the "stratopause." We find it at 50 kilometers or 30 miles on average above sea level. The pressure there is about a millibar. So we see that the troposphere and stratosphere together already account for 99.9% of the mass of the atmosphere. In the stratosphere, temperature increases with height instead of decreasing. We'll see that some incoming solar radiation is absorbed there, and further, harmful ultraviolet radiation is absorbed by oxygen and ozone. The stratosphere to us is the ozone layer.

Next up is the mesosphere, Greek for middle sphere. It's distinguished by temperature resuming its decrease with height. The top of the mesosphere is the mesopause, 85 kilometers, or 55 miles, above sea level. The pressure there is 0.01 millibars, one-hundredth of a millibar. Finally, the thermosphere extends from 55 miles up to where the atmosphere just fades away, and it can live up to its name too. Temperatures there can be thousands of degrees, and yet the thermosphere is also a strangely cold place. The thermosphere helps illustrate the difference between temperature and heat content. There's virtually no mass in the thermosphere, but it's a very deep layer. Great depth with little mass means it is a layer with very, very low density.

I'm sure you've heard the saying "Nature abhors a vacuum." Making one, at least at sea level, is actually pretty hard and, in fact, a good example of a vacuum at sea level is the classic incandescent light bulb. There's a filament in here, and electricity is pushed through the filament in order to heat it up and make light. But air is removed in order to prolong the life of the filament. Even if the air is completely removed and not replaced with anything else, there's still some air remaining in the bulb. There's still some density within this bulb, and the air density in the bulb is still larger than what we would find in the thermosphere.

Suppose you were unfortunate enough to find yourself in the thermosphere, with your back facing the Sun and your belly facing the Earth. You would have the absolutely unique sensation of freezing and frying at the same time. Your back, exposed to undiluted solar radiation, would be very, very hot, and you would find that very deadly. But your chest is hidden from the Sun. So what? If you're in your backyard in your hammock and you're at sea level, your back facing the Sun would certainly feel hotter than your chest facing the ground, but your chest would not freeze because air at sea level is dense enough to carry heat energy around your body from the Sun into the shade. That's not true in the thermosphere. Your chest would freeze.

Let's return to the troposphere. Remember, the troposphere extends 7.5 miles above mean sea level, and we had that huge 140°F temperature drop with height. This is warm air beneath very cold air. It's warm air underneath very, very cold air. If warm air rises and cold air sinks, why doesn't the troposphere turn over? Here's a hint. It's not because the stratosphere is stopping it. It's not because the stratosphere is acting as a lid. The saying "warm air rises and cold air sinks" is not always true. What is true is less dense air rises and more dense air sinks. Less dense air rises and more dense air sinks—that's one of the most important concepts in this course.

We have already seen that density decreases exponentially with height because of air's compressibility squeezing down on itself. The tropopause air is colder. It is colder, but it is also less dense than surface air. In fact, it's a quarter of the density of air at sea level. This is another example where we see temperature, but we need to think density. So why is the troposphere heated from below? Is there any way it could have been different? To understand these questions, we need to know more about how we receive energy from the Sun, and before that, we need to know a lot more about what air is.

So what is air? The first thing we're going to do is we're going to divide air into 2 categories, dry air and water vapor. Dry air is very largely fixed in quantity and well mixed through the atmosphere. Water vapor, in contrast, is extremely variable, in horizontal space, in vertical depth, and in time. So let's look at dry air first. Dry air is 78% nitrogen in the form N_2. That's 2 nitrogen atoms—together they combine a diatomic nitrogen molecule. Nitrogen is removed from the atmosphere by bacteria in soil, and it's returned when plants and animals decay. A further 21% of the dry atmospheric mass is oxygen in the form O_2. Oxygen is removed by plant and animal decay and also by the process of oxidation, in which oxygen combines with other elements. It's added back to the atmosphere through plant photosynthesis.

About 1% of the dry atmospheric mass is argon, one of the nonreactive noble gases. You can think of argon as just lying around, doing nothing. In fact, "argon" comes from the Greek word for lazy. Now, if you were keeping count, that already adds up to 100%, owing to rounding, and it would be very tempting to ignore the minor constituents, but they're not really so minor. We would ignore the other constituents of the dry atmospheric mass at our peril. One of those minor gases is carbon dioxide, CO_2. I want you to think about this analogy. Picture a basketball arena, 25,000 persons gathered together to watch a basketball game, and each individual in that stadium representing a fraction of the Earth's dry atmospheric mass. That means that just shy of 20,000 of the patrons would represent nitrogen's 78%. A little over 5000 would represent oxygen's 21% share. About 250 of the patrons would be argon slackers, probably up there in the cheap seats. In this stadium of 25,000 persons, carbon dioxide's share would represent only 10 persons. You could lose 10 people in a stadium this size and never find them, but these 10 individuals are the players. No players, no game.

Carbon dioxide represents 0.0387% of the dry atmospheric mass, and you know it's rising. That's only 387 molecules out of every million, but carbon dioxide has an outsized importance as a greenhouse gas, so it plays a major role in regulating the temperature of the Earth's surface. Now, I have to admit, my example is a little exaggerated. CO_2 isn't the only greenhouse gas, and indeed, it's not even the most important one. That role goes to water vapor. But what this does show is we cannot presume importance from quantity. So 78% of the dry atmospheric mass is N_2, nitrogen, which has virtually no role in determining the radiative balance of the Earth/atmosphere/ocean system, but this planet would be a different place without that 0.0387% of carbon dioxide. This planet would be a lot different if it had a lot more.

And CO_2 concentrations are rising, due in large part to fossil fuel burning. In fact, CO_2 concentrations in the atmosphere have increased 20% over my lifetime. This plot is called the "Keeling curve," named after Charles David Keeling, the scientist who first started making measurements such as these. The horizontal axis is time in years, and the vertical axis is carbon dioxide concentration in parts per million. Note the saw-tooth nature of this curve. There's an annual cycle superimposed on a clearly distinct upward trend. These data were collected at Mauna Loa, Hawaii, in the Northern Hemisphere.

Which time of year should the atmospheric CO_2 concentration be smallest? Well, let's think. Biological activity of plants removes CO_2 from the air, so therefore atmospheric concentration should be lowest after periods of the highest plant activity, so that means the summer and into the early autumn. This demonstrates that Nature provides an important sink for carbon dioxide, but it shows something more. It shows that the sink is not large enough to compensate for our activity, for what we're doing to increase the CO_2 in the atmosphere. It also shows that without natural sinks, which include the oceans, atmospheric CO_2 concentrations would already be higher.

Other minor constituents in the dry atmospheric mass include methane, CH_4, which is about 2 parts per million. It's added to the atmosphere by 6 distinct sources. First, swamps, rice paddies, landfills, and areas like that. Second, methane is added by ruminants, like cattle and sheep. Third, insects like termites, and also biomass burning, is fourth. Methane is contributed by the oceans as well, and the Earth's interior via volcanic activity. Methane is highly reactive, and it is removed by oxidizers, but the importance of methane is this. It's roughly 25 times more potent per unit weight than carbon dioxide as a greenhouse gas. Methane is also increasing over time, due to anthropogenic activity, and you can see in this plot that it also has a seasonal cycle.

The next minor constituent is nitrous oxide, N_2O, about 300 parts per billion. It's known as "laughing gas," but it's no laughing matter. It's about 300 times more potent a greenhouse gas by unit weight than carbon dioxide. Nitrous oxide is produced in the soil by bacteria, and destroyed by sunlight, principally in the stratosphere. There are anthropogenic sources of this gas and concentrations have been rising.

Our next minor constituent is ozone, O_3, 3 oxygen atoms together to produce an ozone molecule. Ozone represents about 40 parts per billion. Near the surface, it is a dangerous pollutant that damages plants and contributes to smog, and ozone literally eats rubber. If you have windshield wipers on your car and you live in a polluted area, you may notice that they don't seem to last too long, and ozone is part of the reason. Ozone is created by lightning and it contributes to the peculiar fresh odor we sometimes detect in the vicinity of thunderstorms. In fact, "ozone" comes from the Greek word to smell. Up to now, our dry air constituents have been pretty well mixed through the atmosphere, but ozone is our first and most important major exception. The vast majority of ozone is located in the stratosphere, where local concentrations are 250 times larger than they are in the entire atmosphere as a whole.

In the stratosphere, ozone is produced after oxygen molecules absorb incoming solar radiation, and it's destroyed on reaction with atomic oxygen. Ozone itself also absorbs harmful radiation that would otherwise survive to reach the ground. Let's look at the natural stratospheric ozone cycle. It looks something like this. Consider an oxygen molecule, that's O_2, 2 oxygen atoms, and it absorbs ultraviolet radiation. The act of absorbing the ultraviolet radiation causes the molecule to split apart into its individual oxygen atoms, but oxygen atoms don't last very long in the atmosphere. Oxygen is far too reactive and an oxygen atom will find a nearby O_2 molecule to react with, and that makes O_3, which is ozone. Next, ozone also absorbs ultraviolet, which splits it apart. Notice in this process that ozone is both created and destroyed, but there's no net loss. Also, 2 ultraviolet absorption events have occurred, one with oxygen, one with ozone, and that's harmful radiation that doesn't reach the ground.

Now, you've undoubtedly heard of the ozone hole, which is an area of depleted stratospheric ozone, residing over the South Pole. It has a very strong annual cycle, and is most pronounced in October, which is spring in the Southern Hemisphere. The ozone hole is a natural phenomenon, but it's one that has become much more severe in recent years. The peculiar meteorology of the Antarctic atmosphere has made ozone loss far more pronounced there, and in fact, this Southern Hemisphere problem has primarily a Northern Hemisphere cause because, to a large degree, it came out of a spray can.

The culprit in the excessive ozone hole expansion includes something called "chlorofluorocarbons," which I'll call "CFCs." CFCs are very rare in the atmosphere, amounting to only 1 to 2 parts per 10 billion, but CFCs play a dual role as greenhouse molecules and also destroyers of stratospheric ozone. CFCs have anthropogenic sources. They were used as propellants in spray cans, but also coolants in refrigerators and in air conditioners, as solvents and in fire extinguishers. A CFC you've probably heard of is Freon. The problem is CFCs are chemically inert, which leads to them having very, very long lifetimes in the atmosphere. This long lifetime allows CFCs released at the surface to spread globally and loft far upwards, reaching the stratosphere.

In particular, they can work their way to the Southern Hemisphere, where they become trapped in an intense vortex of winds that encircles Antarctica during its long and sunless winter season. When sunlight returns to the region in the Antarctic spring, intense ultraviolet radiation from the Sun can break the CFC molecules apart, liberating the chlorine atoms. Ultimately, and it's more complicated than this, and we'll discuss it, but ultimately, these chlorine atoms become involved in reactions that destroy ozone, but also restore the free chlorine back to the atmosphere, leaving those chlorine atoms free to destroy even more ozone.

The basic chemistry looks like this. First, free chlorine reacts with ozone, creating chlorine oxide, CLO, and molecular oxygen, but then chlorine oxide encounters a free oxygen atom. Remember, those oxygen atoms were just floating around, and when they encounter one, it gives up its oxygen to the oxygen atom. So we see that the free chlorine is liberated and lives yet again to attack yet another ozone molecule. This is Pac-Man in the stratosphere. Now, eventually, the chlorine atom is removed from the atmosphere, but before it has been, it can destroy many, many thousands of ozone molecules.

Let's take a look at what the ozone hole looks like. Here's how the ozone hole appeared on a single day, September 30, 2008, as seen by NASA satellites. We're looking down over the South Pole and we can see the continent of Antarctica. The colored field we're looking at is total ozone in a vertical column, but we already know that virtually all of that is in the stratosphere, so we're really seeing stratospheric ozone. The blue colors in this image indicate very low concentrations, and we see the ozone hole has engulfed all of Antarctica and is reaching out towards South America.

Next, let's look at how the ozone hole evolved within a single year, from the last half of 2008. Our loop has started in July and we see the ozone hole developed by August, the end of Southern Hemisphere winter. Then it gets very big in September, and very big in October, which is their spring. By year end, it's gone, which is summer in the Southern Hemisphere. Note also a clockwise circulation around Antarctic, which we'll come back to. Some interesting geometric shapes come and go as well in the ozone hole at the edges. The ozone hole was first detected in 1979 by scientists of the British Antarctic Survey. They had been making measurements of stratospheric ozone from Antarctica since 1957. This plot shows the Antarctic ozone concentration in October over the years. The depths of the hole grew rapidly after 1979, and the hole has generally been growing larger in horizontal size as well, reaching a record extent in 2006.

The next animation tracks the evolution of the ozone hole over the years. This loop starts in 1979, and we're seeing October ozone concentrations every year during Antarctic spring. We see very clearly there's considerable inter-annual variability. There's great variation among the years, but the trend towards larger and deeper ozone holes, indicated by the preponderance of blue in the fields, is clear. Here's a very good question: Why haven't CFCs destroyed stratospheric ozone over the entire planet? Why is the destruction limited to one place, the South Pole, and one season, the austral spring? The short answer is we got lucky.

The chlorine liberated from the CFCs doesn't stay free very long before getting incorporated into other stable forms, including hydrogen chloride and chlorine nitrate. Under normal circumstances, these compounds can hold onto the chlorine that they acquired from the decomposed chlorofluorocarbons. Remember, you saw a prominent clockwise circulation in those animations. You were seeing the South Polar stratospheric vortex, a huge ring of fast-moving air, encircling the South Pole. That air is extremely, bitterly, incredibly cold, and in that cold vortex, clouds of ice and clouds of nitric acid can form. They may look very pretty, but research has demonstrated that they're deadly to ozone.

They're deadly to ozone because these clouds can play host to chemical reactions that make hydrogen chloride and chlorine nitrate, usually stable compounds, give up their chlorine to form diatomic chlorine gas, Cl_2. Then when sunlight returns in spring, ultraviolet radiation is able to break these Cl_2 molecules up as well, again liberating the chlorine atoms. It is these free chlorine atoms that are involved in the destruction of the ozone, leading to the Antarctic ozone hole. Boy, this is pretty complicated. In fact, I hate to tell you, but the chemistry is even more complicated than that because it involves other chemicals, notably bromine and nitrogen oxide from natural and manmade sources, and these also contribute to ozone destruction.

Indeed, the 1995 Nobel Prize for Chemistry went to 3 scientists for explaining stratospheric ozone processes, including the role of CFCs in their destruction. In 1987, the Montreal protocol, an international treaty dedicated to phasing out the production of ozone-depleting substances like CFCs, was initiated and was one of the most successful international treaties ever. The good news is that new substances that don't deplete stratospheric ozone have been brought into widespread use. But there's bad news too—some of those substitutes are themselves very powerful greenhouse gases, making a little bit of their pollution go a very, very long way.

Now, let's turn to water vapor. Water vapor is literally the fuel of thunderstorms and hurricanes. Water vapor represents zero to 4% of the total atmospheric mass, but is extremely variable in space and time. Water vapor is concentrated near the Earth's surface and the lower troposphere, and there are 3 reasons for that. First, water has a surface source, like soil, plants, and surface water, such as rivers, lakes, and obviously the oceans. Second, an efficient mechanism for returning water to its surface origin exists. I'm talking about precipitation. Third, and most importantly, the ability of air to hold water vapor is a very strong function of temperature.

The ability of air to hold water vapor is a very strong function of temperature. I will discuss an important caveat regarding that statement later, but for now, let's know that warm air can hold much more water vapor than cold air. At sea level, 90°F air can hold more than 7 times more vapor than air at freezing, and as we have already seen, the temperature decreases rapidly with height in the lower atmosphere. This is one reason why most of the water vapor is located near the ground because higher up, it's too cold to permit much vapor to accumulate.

Let's talk a little bit about the early atmosphere. Earth didn't always have the atmosphere we have today. When the Earth was still cooling, it probably had an atmosphere of hydrogen and helium. Those are the 2 most abundant constituents of the universe. But Earth's mass is not great enough, its gravity is not strong enough, to retain those elements, and that early atmosphere was lost to space. Subsequently, Earth developed an atmosphere that had significantly more CO_2 than our present atmosphere, and that atmosphere had plenty of water vapor, nitrogen, sulfur, and ammonia, but the distinctive thing was that that atmosphere had virtually no oxygen. That atmosphere was produced by volcanic activity and crustal outgassing.

We should wonder where all that CO_2 went. Actually, all that CO_2 is largely bound up in rocks, especially carbonates such as limestone. Where did our oxygen come from? In our atmosphere, the first free oxygen probably resulted from photodissociation of water vapor by intense solar radiation. Radiation came in and blasted H_2O molecules apart, liberating the oxygen into the atmosphere, but this is a very, very slow process, so actually, significant oxygen concentrations had to await the evolution of photosynthesis, which converts CO_2 and water into sugar, as well as oxygen. Even then, oxygen concentration increases in the atmosphere were slow, limited by oxygen's highly reactive nature. So a large amount of the oxygen produced by plants and bacteria didn't get into the atmosphere or didn't stay there very long because they became bound up with iron in the soils.

Let's summarize what we've discussed today. The atmosphere has 4 layers. Most important for us is the troposphere, our weather sphere, and the stratosphere, which hosts the ozone layer. In the troposphere, temperature decreases very, very quickly with height. We saw a temperature drop of 140°F over a vertical distance of only 7 to 8 miles, but the troposphere doesn't overturn. The cold air at the tropopause is also much less dense than the air at sea level. Less dense air rises, more dense air sinks. That is the statement that is always true, but what is air?

Air is mainly N_2 and O_2, nitrogen and oxygen, and a little bit of argon. The rest are just minor constituents that don't even add up to 1%, but we saw that those minor constituents are also very, very important. One of those trace gases, CO_2, is a greenhouse gas, and it's increasing in concentration over time. Other even rarer gases are even more potent at regulating Earth's surface temperature, including methane and nitrous oxide, because they're greenhouse gases as well, and they represent the thermostat of the Earth/atmosphere/ocean system. We saw that ozone protects us by absorbing harmful radiation that otherwise would have reached the surface. Manmade chemicals, like the now banned CFCs, destroyed stratospheric ozone, especially in the infamous ozone hole over the South Pole.

So what is this greenhouse effect I've been talking about anyway? What is radiation? What's the difference between the radiation produced by the Sun and the radiation produced by our Earth? We'll find out in our next lecture.

Lecture Four
Radiation and the Greenhouse Effect

Scope:

In this lecture, we'll see that all objects emit radiation, but the amount and type of radiation depend strongly on temperature. We'll also see what solar radiation is made of and why the radiation produced by the Earth is different. Finally, we'll discuss absorption and the greenhouse effect. As we go through this lecture, think about the following questions: What does color tell us about temperature and when? Why can't we see at night? Why isn't the Earth's surface a frozen lump of ice?

Outline

I. Let's begin with the electromagnetic spectrum.

   A. The electromagnetic spectrum consists of radiation we can see (visible light), radiation we can feel (infrared), radiation we can exploit (microwaves for cooking and communication), and radiation we can largely do without (X-rays, gamma rays, and much of the ultraviolet).

   B. The electromagnetic spectrum spans an enormous range of wavelengths. For this course, our focus is on the ultraviolet, the visible, and the near and far infrared.

II. We can identify 4 fundamental facts concerning radiation.

   A. All objects emit radiation.

   B. The total amount of radiative energy emitted is a very strong function of temperature.

   C. All objects radiate energy at all wavelengths of the electromagnetic spectrum.

   D. Objects radiate more energy at some wavelengths than others.

III. Knowing that both the amount and kind of radiation produced depend on temperature leads us to 2 more laws.

   A. Planck's law tells us how much of each kind of radiation an object produces. Wien's law shows us which radiation wavelength is produced the most.

B. For an object with about the same temperature as the Sun's outer surface, the Planck curve looks like a bell curve; the total radiative energy is the area beneath the curve, with the peak in the visible portion of the spectrum.

C. For a cooler object, the area beneath the curve is smaller, and the peak is shifted to the right, to longer wavelengths of visible light.

D. According to Wien's law, the wavelength of maximum emission is inversely proportional to temperature.

E. The Earth's surface typically produces negligible amounts of radiation at visible and ultraviolet wavelengths, which is why we can't see at night.

IV. Let's now turn to emission and absorption.

    A. We see the Planck curves for the Sun and the Earth, plotted on the same scale. Even though the Sun is much hotter, the curves have equal height. The explanation for this is twofold.

        1. First, solar radiation is spread throughout space.

        2. Second, our "radiation budget" is balanced. The energy we receive from the Sun and the energy we lose to the cold of space are equal. This is important for thermal equilibrium.

    B. Note that there is virtually no overlap between these 2 curves, a fact which leads to the greenhouse effect.

V. What is the fate of radiation, whether it's shortwave from the Sun or longwave from the Earth?

    A. Radiation can be reflected back to its origin, it can be scattered in all directions, or it can be absorbed. The only way radiation can change the temperature of an object is through absorption.

    B. Atoms and molecules absorb wavelengths for which they have a particular affinity. Objects that absorb everything are called blackbodies.

    C. The Earth's surface is very nearly a blackbody, while atmospheric gases tend to be very selective absorbers.

VI. A diagram of atmospheric absorption introduces us to the greenhouse effect.

   A. Absorption of incoming solar radiation is limited but important. Much of that radiation survives to be absorbed by the ground. That radiation is then reradiated upward at longer wavelengths. Much of that radiation, though, is absorbed by water vapor and carbon dioxide on the way out.

   B. Our primary greenhouse gases are very selective absorbers, and this differential absorption represents the greenhouse effect. If we remove these gases from our atmosphere, the Earth would be a much colder place.

VII. Let's look at the atmosphere's greenhouse effect in a different way.

   A. Picture sunlight streaming through the atmosphere. Visible, ultraviolet, and near infrared are largely ignored by the finicky atmosphere but absorbed by the ground, warming the Earth.

   B. The Earth radiates longer wavelengths. Some of the outgoing radiation escapes to space, but some is absorbed by the greenhouse gases.

   C. The greenhouse gases themselves radiate in all directions, including downward.

   D. The ground absorbs that radiation and, in turn, must emit. Some of that radiation escapes into space, and some is absorbed by the greenhouse gases.

   E. Each step in the process actually involves less energy. We're moving toward equilibrium, but the temperature of that equilibrium is much higher than if none of the extra greenhouse action had taken place.

   F. The presence of greenhouse gases has made the Earth much warmer than it otherwise would have been. They serve as our thermostat; hence our concern about their quantities.

Suggested Reading:

There is no suggested reading for this lecture.

Questions to Consider:

1. The Sun, as seen from space, is a yellow-white star. Of course, the Sun generates all colors of light, but yellow predominates. As the Sun ages and cools, will its light become more reddish or more bluish? Why?

2. Consider a conventional incandescent light bulb with clear glass that is connected to a source of electricity controlled via a dimmer switch. Turn on the light and slowly rotate the switch, allowing more current to pass through the circuit. At first, the bulb's filament glows a deep red, then a bright orange, then an even brighter yellow before glowing very bright white. Explain this color-shifting phenomenon.

3. Compared to Mercury, Venus is located almost twice as far from the Sun, and receives only a fraction of the solar radiation. Yet, its surface temperatures are hotter than Mercury's, why?

Lecture Four—Transcript
Radiation and the Greenhouse Effect

Welcome back to Meteorology, our survey of the wonders of the weather. Our lecture today concerns energy and the greenhouse effect, but first, let me review what we've seen in previous lectures. We've seen that pressure decreases quickly with height, but the vertical variation of temperature is much more complex. The atmosphere has 4 layers, only 2 of which are actually important to us. The troposphere, or weather sphere occupies, on average, the lowest 12 kilometers, or 7.5 miles, extending above sea level. Temperature decreases with height very rapidly in this region, but it does not overturn because density also decreases very rapidly. Tropospheric temperature decreases with height because it's heated from below, despite the fact that the energy for the Earth/atmosphere/ocean system is coming down to us from the Sun.

The temperature decrease with height ceases in the stratosphere, the next layer up, because oxygen and ozone are there, intercepting some of the incoming solar radiation. In this particular lecture, we'll see that all objects emit radiation, but the amount and type of radiation depends very strongly on temperature. We'll see what solar radiation is made of, and why the radiation produced by the Earth is so very different. We'll discuss absorption and see that some objects are indiscriminate absorbers, while others are very selective. We'll discuss the greenhouse effect. Some questions to think about as we go through this lecture: What does color tell us about temperature and when? Why can't we see at night? Why isn't the Earth's surface a frozen lump of ice? All those questions involve radiation.

In order to discuss radiation, we need to introduce ourselves to the electromagnetic spectrum. The electromagnetic spectrum consists of radiation we can see (visible light, the colors of the rainbow), radiation we can feel (the infrared), radiation we can exploit (microwaves for cooking and communication, radio and TV, and weather radar), and radiation we can largely do without, such as X-rays, gamma rays, and much of the ultraviolet. Radiation travels as waves, and waves are characterized by wavelength, the distance between crest to crest of a wave or from trough to trough. In fact, the electromagnetic spectrum spans an enormous range of wavelengths. Gamma and X-rays are about a wavelength of a billionth of a meter.

Ultraviolet rays are about 10 millionths of a meter. That means you can put 4 million of them in an inch.

Ultraviolet is also about a tenth of a micron, introducing a very useful unit to use in radiation. A micron is a millionth of a meter. Visible light is the range from 0.4 to 0.7 microns, or roughly $2/100,000^{ths}$ of an inch. We divide up the infrared part of the spectrum into 2 sections, the near infrared, about a micron, and the far infrared, closer to 10 microns, about $4/10,000^{ths}$ of an inch. Beyond that, we have microwaves, 1000 microns, or 0.04 inches. TV and FM radio use wavelengths of about a meter, or 40 inches. Finally, AM radio has a wavelength of about 100 meters, or the distance of a football field.

Our focus in this course is ultraviolet, visible, and the near and far infrared. For visible light, let me take you back to your childhood and remind you of the acronym "ROYGBIV." Those are the colors of visible light, arranged from long wavelength to short—red, orange, yellow, green, blue, indigo, and violet. Next, I'd like to discuss 4 fundamental points concerning radiation. I'll mention the 4 points and then we'll discuss them in turn. First, all objects emit radiation. That means the Sun, the air, the ground, and your hair. Second, the total amount of radiative energy emitted is a very, very strong function of temperature. Third, all objects radiate energy at all wavelengths of the electromagnetic spectrum. Fourth, objects radiate much more energy at some wavelengths than others.

So first, the statement was all objects emit radiation—that's true for all objects with a temperature greater than absolute zero. That narrows it down to everything. Remember, absolute zero is −459°F, −273°C, and zero on the Kelvin absolute temperature scale. For our purposes here, suffice it to say that objects emit radiation because they have a temperature.

Second, the radiative energy emitted very strongly depends on temperature. In fact, the energy is proportional to the fourth power of temperature. We call this the "Stefan-Boltzmann law." If we measure the temperature in Kelvin, we see that if we take an object's temperature and we double it, the radiative energy output increases by a factor of 16 because 16 is 2 raised to the fourth power. Picture a warm summer day. The warm ground is maybe 80°F. That's 300 Kelvin. The temperature of the Sun's outer surface is 6000 Kelvin, or

47

nearabouts. That's 20 times hotter. That means the Sun produces 160,000 times more radiation than the warm ground does.

The third point is that all objects radiate energy at all wavelengths of the electromagnetic spectrum. The Sun, the air, the ground, your hair emit gamma and X-rays, ultraviolet, visible light, infrared, microwaves, and radio waves. Beach sand might be hot at noon, but it certainly doesn't give off visible light, and we sure hope it's not producing much deadly gamma ray radiation either. While that statement is true, it's a little misleading, and that leads us to our fourth point.

Objects radiate much more energy at some wavelengths than others. The Sun produces gamma rays and radio waves, but most of its output is in the visible light between 0.4 and 0.7 microns and nearby wavelengths. Beach sand radiates virtually all of its radiation in the far infrared. Emission of visible light, and microwaves, are finite, but undetectably small. So we've seen that the amount of radiation produced depends on temperature, but also the kind of radiation produced also depends on temperature. So that leads us to 2 more laws, which basically illustrate this last point.

Planck's law tells us how much of each kind of radiation an object products. Wien's law shows us which radiation wavelength is produced the most. So let's look at what is called the "Planck curve" for an object of roughly 6000 Kelvin, about the temperature of the Sun's outer surface. The horizontal axis is wavelength in microns from short to long. The vertical axis is a measure of the relative output of radiation at that wavelength. It looks like a bell curve, with a long tail extending to the right. The total radiative energy is the area beneath this curve.

Now, I'm going to subdivide this curve into our 4 principal types of radiation: ultraviolet, visible, near infrared, and far infrared. As I do that, we see the peak for the Sun is in the visible portion of the spectrum, and in fact, the peak is near to the colors green and yellow. Our Sun is a yellowish star, and its color reflects its temperature. Further, we see that almost half of the solar radiation is in the visible band from 0.4 to 0.7 microns. So 44% of the Sun's output is visible light. Another 37% in the near infrared, 11% underneath the long tail of far IR and beyond, and only 7% is ultraviolet and shorter, but this means that more than half of the Sun's energy is actually invisible to our eyes.

So now let's look at the Planck curve for a cooler object, an object that is cooler by a whole 500 Kelvin. That's a lot colder, but it is still a very, very hot object. We see a couple of things. First, the area beneath the curve is a lot smaller. This illustrates the Stefan-Boltzmann effect. Second, there's less radiation at every wavelength, and it may be hard to see, but the peak of the curve has been shifted to the right. It's still in the visible, but it's been nudged a little bit towards the color red, which are the longer wavelengths of visible light. For progressively colder objects, the peak shift to longer wavelengths is now very obvious. This is Wien's law, which tells us the wavelength of maximum emission is inversely proportional to temperature.

We still have a curve that looks like a Planck curve, but the curve is a lot shallower because the temperature is lower, so there's less energy being produced, and the curve's peak is shifting to longer and longer wavelengths as the temperature gets lower. In fact, the smallest curve I've drawn here is for a red dwarf. That's a star with a temperature of 3500 Kelvin, but that's still 12 times hotter than the Earth's surface. Now, you may be wondering something. Why did I draw the hottest object's curve red and the coldest object's curve blue? In fact, I used blue for the curve for the red dwarf. That doesn't seem to make a lot of sense. The answer is in the world of color temperature, our intuition and our experience mislead us. In the world of color temperature, blue is hot and red is cool. An important point here, though, is that the typical Earth surface produces negligible amounts of radiation at visible and ultraviolet wavelengths, and that's why we can't see at night.

So now, let's turn to emission and absorption, which are 2 very important points, and we're leading up towards the greenhouse effect. What I've drawn are the Planck curves for the Sun and the Earth, plotted on the same scale. Now, the Sun is much hotter, so why do these curves have equal height? There are 2 reasons for that. First, solar radiation is spread throughout space. The Sun may produce 160,000 times more radiation than the Earth does, but its radiation is spread out far and wide. The second and very important point is for thermal equilibrium, we need a radiation budget to be balanced. We need in to equal out, so over a long period of time—not so long as to incorporate climate change, but long enough to smear out the effects of seasons and days and sea breezes and things like that—our radiation budget is balanced, and the energy that we receive from the Sun and the energy that we lose to the cold of space are equal.

Note a really, really important point about these 2 curves now that I've justified them—there's virtually no overlap. This lack of overlap leads to an interesting and powerful consequence: the greenhouse effect. I'm going to call solar radiation "shortwave radiation" because it basically occupies the shorter wavelengths of the electromagnetic spectrum, and my synonym for Earth's radiation will be longwave radiation. The division between the 2 is at about 3 microns, but if you look at the curves, you see there's relatively little radiation there, so we don't need to be very precise with our demarcation. So far, what we've seen is how much radiation there is at various wavelengths, but this does not tell us what happens to the radiation, and that's the topic we turn to now.

What is the fate of radiation, whether it's shortwave from the Sun or longwave from the Earth? Three things can happen to radiation. It can be reflected back to its origin, it can be scattered in all directions, or it can be absorbed. It can be absorbed by the ground, by air, by particles in the air, such as cloud droplets and soot. The only way radiation can change the temperature of an object is through absorption, so let me do an example of reflection and absorption. I have here a heart-shaped object. Well, hearts aren't really shaped like this, and it's red, and to a large degree, hearts are red as well. I can actually argue that this heart-shaped object is anything but red. Radiation of many wavelengths is encountering the ball. In the visible wavelengths, all the colors of the light are falling on this ball, and they are all absorbed, except for red, which is reflected back to our eyes. We would not be seeing this object as red if red were not reflected back to us. The old saying is you are what you eat, and if that's true, this object is not red. A black object absorbs all colors of visible light. A white object reflects all colors equally.

Now, let's talk about absorption. Absorption is a very complex subject. Objects, by which I mean atoms and molecules, absorb wavelengths for which they have a particular affinity. Objects that absorb everything, that have an affinity for everything, are called "blackbodies." Blackbodies don't actually have to be colored black, but that helps. The ultimate example of a blackbody is blacktop asphalt. That is a pig that eats anything. Generally, the Earth's surface, dirt, is very nearly a blackbody, and we treat it as such. In contrast, and a very important contrast, atmospheric gases tend to be very selective absorbers. They're finicky eaters. Nitrogen, 78% of the Earth's dry atmospheric mass, absorbs almost nothing. Ozone, O_3, absorbs a lot of ultraviolet, some far infrared, and a wee bit more.

The flip side of absorption is emission. Objects that absorb must also emit. Remember, all objects emit radiation. Absorption depends on affinity. Emission is determined by temperature. Absorption depends on affinity, and emission depends on temperature. Let's consider ozone as an example. Ozone emits in the far infrared, owing to its relatively cool temperature, but it absorbs ultraviolet coming in from the Sun, so the recipe here is a simple one—UV in, far infrared out. What I'm showing you now is a very, very hideous diagram. It is a diagram of atmospheric absorption. The horizontal axis is again wavelength in microns. The vertical axis is now the fraction of radiation that is absorbed by the atmosphere. If the line is very high up on the curve, then that represents very large absorption affinity. So the vertical axis is absorption affinity, and we see a very, very complex figure. This reflects the internal dynamics of atoms and molecules.

Now, when I show this plot in my class, I have a fantasy about this. You remember the movie *Dead Poet's Society*, where Mr. Keating tells his students to rip a page out of their textbook? I keep on wanting to tell my students to take this figure, the atmospheric absorption, and rip it out of their textbooks, and I don't do that. I don't do that because it would ruin the resale value of their textbooks. But this is a very important curve, and so we're going to actually look at this curve step by step, but we're going to keep in mind Einstein's famous admonition, to make things as simple as possible, but no simpler.

So let's build this picture step by step. For the first part, we're going to look at the shortest wavelengths of sunlight, the ultraviolet radiation wavelengths less than 0.4 microns. The simplified absorption for this portion of the spectrum shows that absorption is generally very high. We've pegged it up there close to 100%. That absorption is accomplished by oxygen and ozone in the stratosphere. That's why the stratosphere exists and why it is relatively warm. The ozone hole would let some of these wavelengths pass down to the ground, but notice the absorption curve dips downward before we reach the far end of the ultraviolet. The very longest wavelengths of ultraviolet do reach the ground. They survive to reach the ground everywhere. They reach us. I call this portion of the curve "Sunburn Alley" because these are the radiation wavelengths of ultraviolet that burn our skin. Now, this is not an official term, so don't go telling people about Sunburn Alley.

Let's move onto the next part of the electromagnetic spectrum, visible light. In this portion, we see there is very little absorption of visible light because the absorption curve is very far down at the bottom, so very little absorption affinity. If you look closely, there actually is a little bit, and it's in the longer wavelengths of visible light, the colors orange and red. We'll see later on in this course if this absorption actually means anything, but let's finish the shortwave radiation going down. What's largely left is mainly near infrared, and some far infrared as well, and we see we're back to significant absorption. This is absorbed primarily by water vapor.

Next, let's compare what's being absorbed by the atmosphere and what the Sun is producing, so now I have the curves superimposed on top of each other. You see that the Sun's emission and the atmosphere's absorption curves resemble each other, except they're flipped. There's a very important implication to this. The atmosphere absorbs best what the Sun makes the least of. The atmosphere absorbs best what the Sun makes the least of—half of the Sun's radiation is almost totally ignored. Much of that can pass right through air as if it weren't even there, and it reaches the ground, which will absorb anything, and that warms the Earth.

We're halfway done. Now we're dealing with radiation emitted by the cool Earth as a result of absorbing all that nice visible light. These are longer wavelengths, the far infrared. I've drawn the bell curve of the Earth's Planck curve. This is another bell curve, and it's actually starting to look pretty normal. Let's divide Earth's upper radiation into 3 sections as well, but remember, it's all far infrared. In the first section, we see that absorption is almost 100%. The curve is way up there at the top of the plot. This radiation is being absorbed by water vapor and carbon dioxide. This is outgoing radiation being absorbed and reemitted, some of that being sent back down. This is part of the greenhouse effect.

In the middle section of the curve, emission is large, but absorption is relatively small. We call this the "atmospheric window." Between 7 and 11 microns, much of the radiation emitted at these wavelengths manages to escape to space and isn't recycled into the greenhouse effect. Actually, this picture was a little too simple. Absorption in the 7 to 11-micron band actually looks more like this. There's a smudge, as it were, on the atmospheric window, where ozone does some absorption. I call this the "ozone tonsil" because to me, it looks like one. You may have to stand on your head to see why. Again, that's my term. It's not a real term, but it illustrates why ozone is a greenhouse gas.

Finally, for our last section, the very longest wavelengths of the far infrared, there's less emission, but also lots of absorption. Again, it's water vapor and carbon dioxide. So now, let's put it altogether. Our absorption spectrum: absorption of incoming solar radiation—limited, but important. Much of that radiation survives to be absorbed by the ground. That radiation is then reradiated upwards at longer wavelengths, owing to the cooler temperatures of the Earth. A lot of that radiation, though, is absorbed on the way out, especially by water vapor and carbon dioxide. Our primary greenhouse gases are very selective absorbers, and that is the greenhouse effect. If we remove these gases from our atmosphere, the Earth's atmosphere and the Earth would be a much colder place.

Let's look at the atmosphere's greenhouse effect in a slightly different way. I'm working towards a question: Why doesn't the greenhouse effect run away to catastrophe? I'm working towards an analogy of greenhouse gases acting as a blanket for the Earth's surface. Some people really hate the blanket analogy, but they're joyless people and they're not here anyway, so let's press on. Picture sunlight streaming through the atmosphere. Visible, ultraviolet, near infrared, largely ignored by the finicky atmosphere, but absorbed by the ground and it warms it up. The ground is a blackbody. It'll eat anything.

The Earth also radiates longer wavelengths, reflecting its cooler temperature. This is not the greenhouse effect, at least not yet. Some of the outgoing radiation escapes to space, but some is absorbed by greenhouse gases, those same selective absorbers that wouldn't touch the short wavelength radiation raining down, but they sure like the long wavelength stuff going up. The greenhouse gases themselves radiate in all directions, including down, down towards the ground, where it is absorbed by our old blackbody friend, the ground. It'll eat anything, and as a result, it warms up more.

If you absorb, you must emit, so the ground emits radiation again. Some of this escapes to space. Some is absorbed by our greenhouse gases yet again. What is absorbed again must be reemitted, and some of that is sent towards the ground, which warms up even more. So why doesn't this process run away? Did you notice? At each step, I said "some part of," "some fraction of," each step actually involves less energy. We're moving towards equilibrium, where in equals out, but the temperature of that equilibrium is a lot higher than if none of this extra greenhouse action had taken place.

So let me do my blanket analogy. You're in bed. It's cold. You're losing heat to the cold room. You're losing far infrared. You're also absorbing far infrared from the room, but you're not very happy with the balance you've struck. In equals out, but you're colder than you would want to be, so you toss on a blanket. The blanket absorbs part of your outbound radiation, and some of that radiation, which is absorbed by the blanket, is reemitted back to you, increasing your temperature. Now you feel better. You feel warmer. Since the amount of radiation you emit depends on your temperature, depends on the fourth power of your temperature, you're actually radiating away more energy than ever, but you're also gaining more as well because you're gaining energy not only from the room, but also from the blanket you put on. You have reached a new equilibrium temperature. Your radiation budget is still balanced. In equals out, but you're at a higher temperature. Note that your temperature didn't run away and you didn't melt.

The presence of greenhouse gases has made the Earth much warmer than it otherwise would have been. Earth's average surface temperature, averaged equator to pole, winter to summer, land to sea is about 60°F. Without the greenhouse effect, that average temperature would be 60°F lower. The average would be zero. The Earth's surface would likely be frozen everywhere, including the tropics. The reason the Earth is not a big snowball in space is because of greenhouse gases, carbon dioxide, ozone, methane, nitrous oxide, and especially water vapor, the most important greenhouse gas of all. Greenhouse gases are our thermostat, and hence our concern about their quantities and the future of this.

So let me summarize. The electromagnetic spectrum encompasses a gigantic variety of waves and wavelengths. It's not limited to ultraviolet, visible, and infrared near and far, but those are the portions that we're interested in. Objects emit radiation based on their temperature. The cooler the object, the longer the wavelengths of radiation that are produced. Objects absorb radiation based on their affinity. Some objects absorb everything. We call them "blackbodies." Others are very finicky and selective absorbers. Most air constituents do not absorb much of the short wavelength radiation the Sun produces. A significant and important exception is ozone, which absorbs ultraviolet radiation in the stratosphere, and also molecular oxygen participates in that as well.

The cooler Earth produces longer wavelengths, preferred by finicky gases like water vapor, carbon dioxide, and other minor constituents. This differential absorption represents the greenhouse effect. The very same gases that largely ignored short wavelength radiation on the way down absorb at least some of Earth's longwave radiation welling upwards. Since objects that absorb must also emit, and emit in all directions, some of that intercepted energy is returned to the surface. This makes the Earth's surface warmer than it would have been without the greenhouse effect. In fact, it makes the Earth habitable. The greenhouse effect is not something that's coming in the future. It's here, so we can be also.

Let me do a quick look ahead to the next lecture. We've seen that Earth receives radiation from the Sun, but it is not spread equally across the planet. Variations of solar radiation with time will make the seasons. Variations with latitude will drive the atmospheric circulation. Nature strives mightily to smooth out these differences. There are 2 other ways that Nature has for moving heat energy around: conduction and convection. Only one of them works very well, and we'll find out which one next time.

Lecture Five
Sphericity, Conduction, and Convection

Scope:

In the last lecture, we learned that the potential for life on Earth is due to the atmospheric greenhouse effect. That involved the absorption and emission of radiation, 1 of the 3 ways that heat can be moved throughout the universe. Before we continue in this lecture, think about these questions: What causes the seasons? Why does a metal spoon feel cooler than a drinking glass when both have the same temperature? Why does beach sand get so hot on a sunny summer day, and why is that same sand so cold just before sunrise?

Outline

I. Let's review what we've learned so far.

  A. Temperature differences make pressure differences. Pressure differences drive winds. The intent of winds is to reduce the temperature differences that gave rise to the pressure differences.

  B. Two major sources of temperature difference are unequal heating of the Earth's surface and seasonal variation. The temperature difference is relatively smaller in the summer than the winter, owing to obliquity (the tilt of the Earth's axis), but at no time is the pole warmer than the equator.

  C. The average equator-to-pole temperature difference during wintertime is almost 100°F. It seems, then, that nature isn't very efficient in reducing temperature differences through winds.

II. Let's consider 2 other mechanisms of heat transfer: conduction and convection, both of which require a medium.

  A. Heat is a flow of energy between objects.

  B. Conduction is heat transfer by direct atomic contact. It operates in one direction, from warm to cold. Conduction involves transferring microscopic kinetic energy, as measured by temperature, from an object with more to an object with less.

C. Objects differ in their ability to conduct heat. A metal spoon, for example, feels cooler than a drinking glass, even if they're both at room temperature, because metal is a good conductor of heat.

D. You're perfectly comfortable in 70°F air, but if you were immersed in water at that same temperature, you would feel very cold. Why? Water at that temperature is 26 times better at conducting heat than air is, and since the temperature of the water is lower than your body temperature, you're losing heat to the surroundings.

E. Air is a terrible conductor of heat but a good insulator. This is why beach sand can get very hot on a sunny afternoon, while the air a few millimeters above it is much cooler.

III. In thinking about heat transfer, we also need to look at the concept of thermal inertia, that is, resistance to temperature change.

A. Objects with high thermal inertia can absorb a lot of energy without their temperatures rising very much.

B. Note that heat conductivity and thermal inertia are not the same thing. The afternoon sand on the beach feels hot because it is a good conductor, and it gives you its heat very quickly. The sand is hot because its thermal inertia is low.

C. In contrast to the temperature of sand, which has a wide-ranging diurnal cycle, the temperature of the sea surface doesn't change much during the day. Part of the reason for this is that the sunlight warms up the surface of the water, but then that water is circulated internally.

D. If air were a better conductor of heat, the sand on the beach would not get as hot as it does because the air would take heat away from the sand more quickly.

IV. Given that air is not a good conductor, heat transfer in any direction needs some help; that help comes from convection, heat transport by mass fluid motion. In this course, heat transport basically means wind.

A. One way the wind can cause vertical mixing and vertical heat transport occurs as it courses over irregular surfaces. This creates eddies that transport hot air up from the ground and cooler air down to the surface.

B. The same thing can occur as air passes over land that has been irregularly heated.

C. Looking at a temperature profile with height for a windy day, we see that the temperature near the ground is lower than it would be if the air were calm.

D. On the beach, sand's low thermal inertia permits the sand to get cold at night, chilling the air near the ground, and the warmth from above is not conducted downward very quickly. This results in a temperature inversion, or a radiation inversion, when temperature increases with height.

E. Under what conditions are we most likely to get really cold temperatures at the ground? There are 4: low humidity, a clear rather than cloudy sky, low thermal inertia of the surface, and calm winds.

V. Temperature differences play a critical role in meteorology.

A. One way to create temperature differences is to provide surfaces that differ in some critical aspect. Suppose they receive more or less solar radiation, owing to Earth's curvature and tilt. They might also differ in how they absorb solar radiation due to differences in reflectants.

B. One way to mitigate temperature differences created by differences in latitude, season, or surface characteristics is to transport heat by mass fluid motion—by convection.

Suggested Reading:
There is no suggested reading for this lecture.

Questions to Consider:
1. Seasons are caused by the fact that the Earth's axis is tilted. Presently, that tilt is 23.5 degrees, but this value varies over a long period of time. When would you expect an ice age to be most likely: When the tilt is greater than the present value, or smaller? Why?

2. Consider a blanket of snow on the ground. It is often noticed that snow melts faster from below than above—resulting in air pockets beneath the crust that causes the snow to crunch when you walk on it. Why?

Lecture Five—Transcript
Sphericity, Conduction, and Convection

Welcome back to Meteorology, our survey of the wonders of the weather. In the last lecture, we learned why we have a chance at having life on Earth—the atmospheric greenhouse effect. That involved the absorption and emission of radiation, one of the 3 ways that heat can be moved around on Earth and through the universe. Before we review, I have a few questions for you. What causes the seasons? Here's a hint: Most Harvard graduates get this one wrong. Why does a metal spoon feel cooler than a drinking glass, when both have the same temperature? When we touch it with our hand, why do we tend to dramatically underestimate the temperature of the sea? Why does beach sand get so hot on a sunny summer day? The answer is surprisingly complex. Why is that same sand so cold just before sunrise?

So here's the story so far. Temperature differences make pressure differences. Pressure differences cause winds. How and why do temperature differences arise? An even more elementary question is how does heat energy move around? We're considering this in a step-by-step process. The primary energy for the Earth/atmosphere/ocean system is the Sun. That energy reaches us through space by radiation. Being hot, the Sun's radiation is concentrated primarily in the shorter wavelengths we've termed shortwave radiation, ultraviolet, visible, and the near infrared. As this radiation passes through our atmosphere, it can be reflected back to space, scattered around, or absorbed. Some radiation in the ultraviolet is absorbed by oxygen and ozone, but for the most part, air is a poor absorber of shortwave radiation, so a substantial fraction of the Sun's energy survives to reach the ground, where it is much more readily absorbed. The ground is nearly a perfect absorber. The ground is a blackbody.

As a consequence, although the Sun is out there in space, to a very large degree, the atmosphere is heated from below, like a pot of water on your stove. This is why the warmest temperatures in the lower atmosphere are found near the ground and decrease swiftly with height. This was the portion of the atmosphere we called the troposphere. The temperature only stopped decreasing with height because of the absorption of ultraviolet in the stratosphere. Without oxygen and ozone to intercept much of that shorter wavelength ultraviolet radiation, the Earth would be a very different place.

In this course, we will see that the variation of temperature with height in the troposphere plays a very important role in the weather. That variation of temperature with height is in part a consequence of differences in how air and earth absorb radiation, and also in part a consequence of how the atmosphere moves heat around. Also crucially important is the fact that radiation is not equally distributed across the Earth's surface. Instead, the tropics receive much more solar radiation than the poles. This creates a temperature difference between equator and pole. Since temperature differences make pressure differences, and pressure differences drive winds, the equator to pole temperature difference is the primary driver of the atmospheric circulation, and it's all a consequence of the fact that the Earth is spherically shaped.

I'm going to do an example with a flashlight, and I'm going to turn the flashlight on. I'm going to point the flashlight towards the table. I'm producing a certain amount of light, and this light is basically striking the table over a relatively small area. This represents the direct radiation that is received by the tropics. But if I tilt the beam of the flashlight, I'm not producing any less light than I did before, but now it's spread over a larger area, so any single piece of this area that is being illuminated is actually receiving less energy. As a result, it's absorbing less energy here at the surface, and this represents the poles.

Since the Earth is spherical, the Sun's rays are less concentrated at the poles. They're making a more acute angle to the ground, and that's the primary reason why the poles are colder. This would be true even if the Earth didn't have an atmosphere, but atmospheric effects compound this problem in direct and indirect ways. While absorption of sunlight by the air is pretty limited, the amount of solar energy that reaches the ground can be reduced by reflection and scattering, so the potential for this loss depends on sunlight's path through the atmosphere. In the tropics, the Sun passes nearly overhead during the day. This decreases the distance sunlight has to travel through the atmosphere, but at the poles, the light makes a much more acute angle with the ground as a result of the fact that the Sun travels lower in the sky. So the path taken through the air is longer, and that increases the chances for losses along the way, reflections and scattering and things like that, making the amount that actually reaches the ground even less than expected.

A further complication arises from the fact that the Earth's axis is tilted. We call this "obliquity." Our present tilt to the Earth's axis is 23.5 degrees from the vertical, but this varies between 22 and 24 degrees over a period of about 41,000 years. Owing to this tilt, both hemispheres spend part of the year inclined towards the Sun and part inclined away, and this causes the seasons. So we're looking at Northern Hemisphere summer. At the North Pole, the Sun's rays are still coming in at an acute angle at the pole, but more directly than they would have if the Earth were not tilted, so the poles are relatively warmer in the summertime. Tilting away from the Sun creates winter. When one hemisphere tilts away, the other is tilting toward, so obviously, seasons will be opposite on either side of the equator. Our winter is the Southern Hemisphere's summer.

But that's not the only complication. The Earth's orbit is also eccentric. Eccentric means that the Earth's orbit is not perfectly circular. There are periods of the year when we're relatively closer or farther away than our mean distance of 93.5 million miles. This is sometimes confused as the reason for the seasons, but keep this in mind: What time of year are we actually closest to the Sun? Actually, it's in January, during the Northern Hemisphere's winter. So eccentricity is a very small complication. Here is the recipe: no tilt, no seasons.

So let's retrench. Temperature differences make pressure differences. Pressure differences drive winds. The intent of winds is to reduce the temperature differences that gave rise to the pressure differences in the first place. One major source of the temperature difference is unequal heating of the Earth's surface, owing to the Earth's round shape. There is a seasonal variation as well. The temperature difference is relatively smaller in the summer than the winter, owing to obliquity, but at no time is the pole ever warmer than the equator. The average winter surface temperature at the North Pole is −30°C, or −22°F. An average year-round temperature in the tropics is about 25°C or 77°F. This makes an equator to pole temperature difference of almost 100°F during wintertime.

If the purpose of winds is to reduce temperature differences, if the purpose of winds is to eliminate imbalances, smooth out inequities, and yet we still have a temperature discrepancy of 100°F between the equator and pole during the winter season, shall we conclude from this that Nature is not very good at what she does? Should we believe that perhaps Nature doesn't abhor extremes as much as I might insist? I say the answer to that is no. Before we convict Nature of slackness or failure, we should consider what the Earth would be like if Nature didn't at least try to address heating imbalances. A very ready example is our immediate neighbor, the Moon. The Moon is also an unequally heated sphere, but the Moon has no atmosphere. There are no winds to move heat energy around.

Our wintertime temperature difference between equator and pole may seem very large at 100°F, but the Moon's equator to pole temperature difference is closer to 500°F. So Nature may not be as successful as we might like, but it has already accomplished a great deal with the atmosphere, its winds, clouds, storms, and all her other tools. Yet it's still worth considering why Nature hasn't done better for us. This leads us to study the obstacles and frustrations that Nature faces.

One of these is the fact that the Earth rotates, but we're getting a little ahead of ourselves. Before that, we need to understand how and why local temperature differences can lead to local winds, and what those winds are able to accomplish. In this lecture, we have 2 more goals: First, to appreciate that other ways of moving heat energy around, besides radiation, also exist, and why they're not equally important. Second, to determine how quickly an object heats up or cools down when exposed to that radiation, since that helps make local temperature differences. So the Sun's energy reaches us through the empty vastness of space via radiation. This is a very efficient means of transporting energy that does not require mass as a medium.

There are 2 other mechanisms of heat transfer we need to consider: conduction and convection. Both require a medium. In our case, we're concerned with air and water, but first, let's consider a question. What is heat? We often think of heat as a property. We say that something that is hot has a lot of heat, but properly speaking, heat is a flow of energy between objects. The property we're thinking of is actually internal energy, and temperature helps us measure that property. We want to move that property around and that is the subject of what we call "heat transfer," a term that ironically still treats heat as a property.

Central to the discussion of heat transfer is a concept called "heat capacity," the ability to hold heat, another legacy of archaic thinking.

I will break free of this past by using a different term for heat capacity in the discussion that follows, but I will likely wind up treating heat as a property like everyone else, which isn't correct, but it certainly seems very sensible to us. So let's talk about conduction. Conduction is heat transfer by direct atomic contact, atoms and molecules touching each other. Conduction operates in one direction, from warm to cold. Recall we discussed that temperature is a measure of microscopic kinetic energy of atoms and molecules, vibration, translation. For us, we can just think about the vibration part of that.

When warmer or more vibrant molecules and atoms come in contact with cooler, less vibrant ones, some microscopic kinetic energy is transferred from the warmer object to the cooler one. So heat is transferred from warm to cold. I find it helpful to picture a billiards table, with balls moving around. This is macroscopic kinetic energy, but what happens on the billiard table when faster and slower balls collide? Does the faster-moving ball speed up as a result of the collision? No. It slows down instead and part of its macroscopic kinetic energy is transferred to the ball that's moving with the smaller speed. So there has been a transfer of kinetic energy from the ball with the excess to the ball with the deficit. Conduction, in a sense, is atomic billiards. Conduction involves transferring not macroscopic kinetic energy, but microscopic kinetic energy, as measured by temperature, from the object with more to the object with less. In this way, conduction moves heat from warm to cold.

The next important point is that objects differ in their ability to conduct heat. I have a metal spoon and a drinking glass. Both of these have been sitting here for some time and they have the same temperature, but if I touch them, I feel a temperature difference. Which feels cooler? The spoon feels cooler than the glass, but they're at the same temperature, so why do they feel different? The reason is that metal is an excellent conductor of heat. The spoon feels cool to our touch because it's taking heat away from our skin quickly. I am warmer than the spoon and the spoon is conducting heat away from me. Materials that are dense and with highly ordered atomic structures, like metal, are generally very good conductors. They're very good at playing this game of atomic billiards, or atomic dominos.

This is why we use metal pots and pans in cooking, to quickly heat the food in the pan by the metal. Iron is good, aluminum is better, and copper and silver are the best heat conductors of all. In contrast, glass is a pretty poor conductor. It takes a lot longer to heat food up in a glass pan. The glass is not conducting the heat to the food as quickly, but food stored in glass will keep its high temperature longer since the glass is slower to conduct that extra heat away. Our old saying is, cook with metal and serve with glass. Slow cookers use glass-lined pans for the same reason. In that case, it does not matter if things actually heat up quickly or not.

Now, let's consider an example using water and air. You're perfectly comfortable in 70°F air, wearing light clothing, but if you were immersed in water at that same temperature, you would feel very cold. In fact, prolonged exposure to that water would kill you. Why? First of all, we need to consider that water at this temperature is 26 times better at conducting heat than air is, and since the temperature of the water is lower than your body temperature, you're losing heat to the surroundings. Your heat loss to 70°F air is pretty slow and you can deal with it because air is not a very good conductor, but the heat loss to 70°F water is much more of a problem because it's a lot faster, even though we should keep in mind that water is still 650 times worse at conducting heat than copper.

The point of this is that air is a terrible conductor of heat. Now, that's not all bad. Air is actually a very good insulator. It's good at stanching heat transfer. It's good at slowing heat loss. In fact, air is the secret to some of our very best insulators. Here's a piece of fiberglass insulation. You probably have this in your attic. What makes fiberglass insulation a great insulator is the fact that there are many air pockets between the fibers of the insulation. If you want to prove that, all you have to do is compress the fiberglass down and make it as thin as possible. Squeeze out the air and close up those air spaces, and you'll find that it does not insulate as well as it did before.

Igloo snow houses are another example of a good insulator. Snow is a terrific insulator because the irregular structure of snowflakes helps create air spaces that trap air. Woolen sweaters also trap air in the fibers of the wool. Cork pads that we use on our kitchen tables—the secret is lots of air pockets. The consequence of air's poor conductivity of heat is this: Air does not carry heat away from a heated surface very quickly. An example is beach sand on a sunny afternoon. The sand can get very hot, but the air even a few millimeters above the sandy surface can be much cooler. The air is not taking away the sand's excess heat very efficiently because air is a lousy conductor.

Then picture what happens if the tide washes up. How quickly does the sand cool off then? Even if the water were the same temperature and density of air, the sand would cool much faster, due in part to water's much larger conductivity of heat. So heat conduction through air is too slow, too inefficient to be a useful way for Nature to move heat around. In fact, if you waited for conduction alone to heat your broth for dinner, you might starve first. Air is a much worse conductor than water is. But in thinking about heat transfer, conductivity isn't the only property we need to be concerned with. The next concept is called "thermal inertia."

Inertia is resistance to change. By analogy, what I will call "thermal inertia" is resistance to temperature change. Objects with large thermal inertia can absorb a lot of energy without their temperatures rising very much. You've probably observed that objects that tend to warm up quickly also cool off very quickly as well, and vice versa. That's thermal inertia at work, resistance to temperature change up or down. This concept is usually called "heat capacity," the ability to contain heat as if it were a property. Let's use sand as an example. Sand has a very low thermal inertia, so back on our sunny, warm afternoon, the sand can get very hot. It got hot during the day because the sand could not resist the temperature rise as it absorbed solar energy from the Sun. Overnight, that same stretch of sandy beach will get very cool. It gets cold at night because it cannot resist the temperature drop, as it loses energy via radiation.

Now, we need to consider a point of potential confusion. Heat conductivity and thermal inertia are not the same thing. This is a pretty subtle point. The afternoon sand feels hot because it is a good conductor, and it's giving you its heat very quickly. The sand is hot because its thermal inertia is low. In fact, if the sand did not have a low thermal inertia, it would not have gotten so hot in the first place. In fact, it might have even felt cold to you instead of hot if its thermal inertia were very high. So now let's contrast sand and sea.

Owing to its low thermal inertia, the temperature of sand can have a very large diurnal cycle. At the same time, the sea surface temperature doesn't change very much during the day. One reason for that is the sea is translucent to sunlight and it has an internal circulation. When the sunlight reaches the sand, you're just heating up the top millimeter or so of sand. Dig a couple of inches beneath the surface of the sand and you'll see that the soil there is much cooler. But when sunlight lands on the sea surface, you're warming up water that is absorbing the solar radiation, but then that water is replaced by other water by an internal circulation, and so actually, you're mixing that heat over a greater depth. As a result, the ocean is a huge reservoir of heat energy, as well as water, and the thermal inertia of air is somewhere in between sand and sea.

Here's a question. On that sunny beach, if air were a better conductor of heat, would the sand get as hot as it does? Let's analyze the situation. It's a warm, sunny day. The sand is a far better absorber of solar radiation than the air above it, so the sand is absorbing energy that the air is not. Is this why the sand gets hot? That's not enough. Further, the sand's thermal inertia is low, so this energy absorption causes its temperature to get high. Now we've finished our answer. Sand is a pretty good conductor and it's willing to share its heat with the cooler air molecules that come bumping on by. But air is a poor conductor, so it does not share its bounty of heat energy acquired from the sand with other air molecules above very quickly.

Let's take a look at a vertical profile of temperature very near the ground. Here is Stick Man, and I'm using Stick Man for a sense of scale. The horizontal axis here is temperature, from cold to warm. The vertical axis is height, from the surface to a little above Stick Man's head. Stick Man can use a few pounds and I can lend him a few, but what this graph shows is that temperature can get very hot near the ground over that sandy surface, but it decreases very quickly with height. The air at Stick Man's head can be quite a bit cooler than the air at his feet, or at his knees. I'm showing 120°F at the ground, and only 90°F at head height.

If air were a better conductor, it would take heat away from the surface more quickly, so the sandy surface would not have been able to get as hot, and the air at Stick Man's head would likely be warmer as well. Now I'm showing you what the temperature profile might look like if air were a better conductor. Notice that the temperature near the ground is now cooler than it was before, but the temperature at the height of Stick Man's head is actually warmer because the heat has been shared more equally through the lowest layer of the atmosphere. Air is not a good conductor, so heat transfer in any direction needs some help, and we call that help "convection," heat transport by mass fluid motion.

"Convection" comes from the Latin word to carry, to convey. Basically, I'm talking about the wind. One way that the wind can cause vertical mixing and vertical heat transport occurs as the wind courses over irregular surfaces, mountains, hills, trees, houses, buildings, things like that. This creates eddies that transport hot air up from the ground and transport cooler air down to the surface. The same can occur as air passes over land that has been irregularly heated, some places getting warmer than others, perhaps because you have concrete, you have grass, and places with different thermal inertias.

Due to convection, a windy day can accomplish what conduction cannot, and that is to make the sandy surface cooler than it otherwise would have been. So now you're looking at a temperature profile with height for a windy day. Notice that, again, the temperature near the ground is lower than it would have been if the air were calm. The temperature at Stick Man's head is higher as a result. At night, sand's low thermal inertia permits the sand to get cold, and this will chill the air near the ground, but the warmth from above is not conducted downward very quickly. In this plot, we're seeing what we call a "temperature inversion," when temperature increases with height. We call this an "inversion" because it's the opposite, or inverse, of what we've come to expect, which is that temperature decreases with height above the ground.

Technically, this is called a "radiation inversion," as it's caused by radiative cooling on the sandy surface, exacerbated by the low thermal inertia of sand. But as we've also seen, it's also exacerbated by air's poor conductivity. If air were able to share its bounty of heat from a few feet above more efficiently, it would not get so cold at the sandy surface at night. But that's for a calm night. When the wind kicks up, again, there's more likely to be vertical mixing, this time leading to relatively warmer temperatures near the ground and cooler temperatures farther aloft. So on a windy night, it's likely to be not quite so cold at the sandy surface, and likely to be a little colder at Stick Man's head.

A few more questions: Under what conditions are we most likely to get really cold temperatures at the ground? I can think of 4 of them. First, when the air is relatively dry, there's less water vapor to trap outgoing longwave radiation, which is part of the greenhouse effect. Other things being equal, less humid nights are cooler than more humid nights. Second, when the sky is clear, no clouds to absorb outgoing longwave radiation and return some of that to the surface, so other things being equal, clear nights are cooler than cloudy nights. Third, when the surface has a low thermal inertia. Sand and concrete will get colder at night than grass and water, other things being equal. Fourth, when the winds are calm because this limits the vertical mixing of air and helps to maintain the surface cooling.

So let's summarize what we've discussed in this lecture so far. We have a large temperature difference between equator and pole because the Earth is spherical. That temperature difference is exaggerated in the winter season because our half of the sphere is tilted away from the Sun at that point in time. Axial tilt causes the seasons. The winter equator to pole temperature difference is 100°F. Without the atmosphere and its circulation, though, that temperature difference would be a heck of a lot larger. There are 2 other ways of moving heat around. Both require a medium. Conduction, atomic billiards; convection, carried by the wind. Conduction is slow, in air at least, because air is a lousy conductor, but that makes air a great insulator.

We examined a concept called "thermal inertia," the resistance to temperature change. Objects that heat up slowly tend to cool off slowly as well. Liquid water has a huge thermal inertia and, as a consequence, our oceans are also huge reservoirs of heat energy, and not just water. Many land surfaces, in contrast—I've been using sand as an example—have low thermal inertia. It heats up fast during the day and it cools off fast at night. So let's look ahead. Temperature differences play a critical role in meteorology. One way to make temperature differences is to provide surfaces that differ in some critical aspect. Suppose they receive more or less solar radiation, owing to Earth's curvature and tilt, such as the equator and the pole, winter and summer.

They can also differ in how they absorb solar radiation, due to differences in reflectants. Consider a shiny surface versus a dark surface. One of the reasons why snow can persist so long is it reflects away so much of the solar energy due to its brightness that would want to melt it. These surfaces could also respond to the energy that they absorb differently, owing to differences in thermal inertia. They also can differ in how they conduct heat away when they're hot, and receive heat when they're not. One way to mitigate temperature differences created by differences in latitude, differences in season, or differences in surface characteristics is to transport heat by mass fluid motion by convection, by the winds. We can create winds and mixing through mechanical and thermal processes, and transfer heat far more efficiently than through atomic billiards.

Temperature differences make pressure differences, and pressure differences drive winds. Here's a question for you. When the winds blow, is heat also forced to move from warm to cold? In the next lecture, we will see that the quintessential local circulation responding to temperature differences, owing to radiation and conduction, is the sea breeze.

Lecture Six
Sea Breezes and Santa Anas

Scope:

As we've seen in recent lectures, temperature differences can arise for many reasons. In this lecture, we'll see how local temperature differences, such as between land and sea, can create a local circulation, the sea breeze. The purpose of these winds is to reduce the temperature differences that gave rise to them. As always, here a couple of questions to think about during the lecture: We've heard that heat flows from warm to cold; how is it, then, that we experience cold winds? Which would be better in eliminating temperature differences: blowing warm air to a cold place or blowing cold air to a warm place?

Outline

I. Let's begin by revisiting pressure.

 A. Pressure largely reflects the weight of overlying air, owing to gravity, and is proportional to mass. Therefore, pressure decreases with height.

 B. Surface pressure is approximately 1000 millibars, and 50% of the atmospheric mass resides between the levels of 1000 and 500 millibars.

 C. We know that pressure decreases with height; note, too, that pressure decreases with height faster in colder air.

II. Let's see how temperature differences make pressure differences.

 A. If we increase the temperature on one side of the space between the 2 isobars, the 500-millibar isobar rises higher above the ground. The layer of air between the 2 isobars is thicker because the temperature is warmer.

 B. Let's pick a point and draw a line along the 500-millibar level, then look toward the colder air. Both isobars are beneath the line. The pressure at this point is lower than 500 millibars.

 C. Looking toward the warmer air, the 500-millibar level is above our line. The pressure is higher than 500 millibars here.

D. We have relatively higher pressure at the same level in the warmer region. We have relatively lower pressure at the same level in the colder region. This is a pressure difference.

E. Pressure differences create winds. Here, the air within the warm region rises up and diverges out of that region. Mass is leaving this relatively warmer region. And since there's less mass in that region, the surface pressure is dropping.

F. The warmer air is moving toward the colder region, adding mass to that region and causing the surface pressure there to rise.

G. This tilts the 1000-millibar isobar, leaving us with relatively higher pressure where it's colder near the surface and relatively lower pressure where it's warmer near the surface. We've created a circulation.

III. Let's apply this model to the sea breeze.

A. The ocean absorbs more solar radiation during the day than the sand does because it's darker.

B. The ocean isn't hotter than the sand because it has an internal circulation and because liquid water has significant thermal inertia.

C. Looking at the sea breeze circulation, we see warm air rising over the heated land and cool air sinking over the cooler ocean. The surface sea breeze is blowing inland from the cooler sea to the warmer land.

D. A topographic map of the city of Los Angeles, CA, shows the increase in the wind from sea to land over the course of a day.

E. If the land surface becomes colder at night, the circulation will reverse, and a land breeze will develop.

IV. How do these breezes work to decrease temperature differences?

A. To answer this question, think of eating hot soup in a cold room. The soup loses heat energy to the room air via conduction, but the air doesn't carry the heat away very efficiently.

B. As a result, the temperature difference between the soup and the air decreases as a function of time. This, in turn, decreases the heat loss because heat loss is proportional to temperature difference.

C. There are at least 2 other mechanisms for helping to reduce the land/sea temperature difference operating in this case: (1) Mixing land and sea air moderates the temperatures of both, reducing the temperature difference from land to sea, and (2) changing the elevation of air changes its temperature.

D. Specifically, air warms in descent. The temperature change is due to volume change alone, without the addition or removal of heat. This is called the dry adiabatic process.

E. In general, rising air cools and sinking air warms up. Thus, the vertical motions in the sea breeze work to decrease the temperature difference that drove the circulation to begin with.

V. The Santa Ana winds of southern California blow dry, hot, and fast.

A. Most Santa Ana events start with cold, dense air, spilling down across the Great Basin of Nevada and Utah. The southward progress toward Los Angeles is partially stopped by a ring of mountains surrounding the Los Angeles basin.

B. When the cold air reaches the mountains, the flow is restricted, but the wind that escapes is blowing at a higher velocity. As it descends into the basin, the air experiences compression and is warmed at a rate of 30°F per mile, or 10°C per kilometer.

VI. Ascending air cools at the same rate, 10°C per kilometer, which is a ratio of 2 numbers: g, the acceleration of gravity, and cp, a measure of thermal inertia.

A. The value g tells us how quickly an object in freefall gains velocity on descent: 32 feet per second squared, or 10 meters per second squared.

B. The value cp is the specific heat of the air at constant pressure. In the metric system, cp is 1004 joules per kilogram.

C. The rate of 10°C per kilometer thus reflects the Earth's mass and composition.

Suggested Reading:

There is no suggested reading for this lecture.

Questions to Consider:

1. Is the dry adiabatic lapse rate also 10°C per kilometer or 30°F per mile on Venus, Mars, or Jupiter? Why or why not?

2. How would the Santa Ana winds be different if you flattened the topography of the western United States?

Lecture Six—Transcript
Sea Breezes and Santa Anas

Welcome back to Meteorology, our survey of the wonders of the weather. As we've seen in recent lectures, temperature differences can arise for many reasons. Different locations receive varying amounts of radiation from the Sun depending on location, time of year, and also time of day. Surfaces vary in how they absorb the radiation they do receive. Some are very good absorbers, like blacktop. Others are not. Others would be even better absorbers if they didn't reflect away the radiation that they do receive, due to brightness, like snow and light sand. Surfaces also vary in how they respond to received energy. The temperature of substances with higher thermal inertia, such as seawater, doesn't change very much as they absorb radiation.

This energy, however acquired, is not shared efficiently with the overlying air. Air is a poor conductor of heat due to its low density and lack of structure at the molecular level. The third mechanism of heat transport is convection, heat transport by the winds. Temperature differences make pressure differences, and pressure differences drive winds. In this lecture, we will see how local temperature differences, such as between land and sea, can create a local circulation, the sea breeze. The purpose of these winds is to reduce, if not eliminate, the temperature differences that gave rise to them. The purpose of these winds is to put themselves out of business.

So a couple of questions to think about as we go through this lecture: We've always heard that heat flows from warm to cold, but we've experienced cold winds as well, so is that statement, in fact, always true? In fact, which would better eliminate temperature differences if you could only accomplish one, to blow warm air to the cold place, or to blow cold air to the warm place? We'll see.

First, let's revisit pressure. Pressure largely reflects the weight of overlying air, owing to gravity, and is proportional to mass. Therefore pressure decreases with height. Remember, surface pressure is approximately 1000 millibars, and 50% of the atmospheric mass resides between the levels where the pressure is 1000 and the pressure is 500 millibars. Picture a vertical cross-section. I've drawn here 2 isobars. Isobars are lines of equal pressure. This comes from Greek, "iso" meaning equal and "bar" you see in the word "barometer." Here are the 1000 and 500-millibar isobars, and we know that half of the atmospheric mass is in between. The 500-millibar level is only 5.5 kilometers, or 3.5 miles, above our heads, on average, so that the 1000 to 500-millibar layer is only 5.5 kilometers, or 3.5 miles, thick.

Use your intuition. I have a layer of air, the 1000 to 500-millibar layer. It contains a certain amount of mass. What if I warm up that air? It wants to expand, right? So what we've now realized is temperature affects thickness. When it's relatively warmer, the 1000 and 500-millibar isobars are farther apart. The 1000 to 500-millibar layer is thicker. By the same token, when the temperature is relatively colder, the 1000 to 500-millibar thickness is thinner. Something that's very important about this as well is that it demonstrates that although pressure decreases with height, pressure decreases with height faster in colder air.

Next, we're going to demonstrate that temperature differences make pressure differences. Let's start with the same thickness everywhere. This implies the temperature is the same, but let's introduce a temperature difference. Let's make it warmer to the right. Now the 500-millibar isobar is relatively higher above the ground because the 1000 to 500-millibar thickness is thicker because the temperature is warmer. Do you see a horizontal pressure difference here? Let's pick a point along the 500-millibar level, and let's look towards the colder and the warmer air at the same height.

If we look towards the colder air, we see that the 1000-millibar isobar is beneath us, and the 500-millibar isobar is beneath us as well. The pressure where we are at this point is lower than 500 millibars. It's lower than when we started. But now let's look towards the warmer air. The 500-millibar level has risen, and if I look towards the warmer air, I see the 1000-millibar level is below me, but the 500-millibar level is above me. So what I see from this is the pressure is higher than 500 millibars here. I have relatively higher pressure at the same level in the warmer column. I have relatively lower pressure at the same level in the colder column. Now I have a pressure difference.

Pressure differences want to make winds. Nature wants to move the mass from high to low pressure, but we're not done yet. Where does the air blowing out of the warm column come from? It's rising up within the warm column, and air is diverging out of that column. Do you see that? So what that means is that mass is leaving this relatively warmer column. I made this air warmer. I made the air thicker, but now the wind is blowing and the air is leaving this column. Since there's less mass in this column now, the surface pressure is dropping. Now, where has that air gone? The air is moving towards the colder column. It's converging into the colder column, adding mass to that column, and that causes the surface pressure there to rise.

So now the 1000-millibar isobar is tilted as well, and we have relatively higher pressure where it's colder near the surface, and relatively lower pressure where it's warmer near the surface, and we've made a circulation. So let's review our circulation very quickly. It all started with a horizontal temperature difference. That made a pressure difference somewhere above the ground. We looked near the 500-millibar level, a few miles up. That made a wind. The wind caused air to move out of one column and into another. The column losing mass had its surface pressure drop. There's less mass pushing down and therefore less pressure at the surface. The column gaining mass saw its surface pressure rise. That made a surface pressure difference that didn't exist before, and a wind that closed the circulation loop.

Note that, at least at the surface, the wind is blowing from the colder to the warmer place. Now we're going to apply this to the sea breeze. We start with ocean and land. Which surface absorbs more solar radiation during the day? Actually, the answer may surprise you. The ocean does. The ocean absorbs more solar radiation during the day because it's darker. The relatively light color of the sand actually reflects away a fairly large fraction of the sunlight. When the Sun is high in the sky, the ocean surface is quite dark. What is not absorbed can't change the temperature. You may be wondering then why the ocean isn't hotter than the land, since it is absorbing more solar radiation. We've seen already 2 reasons for that. First, ocean water has an internal circulation. It's moving the warmed water around. With the sand, you're just heating the very, very top. Second, liquid water has a huge thermal inertia. It resists temperature change, up or down.

Here's the sea breeze circulation. We have warm air rising over the heated land, cool air sinking over the cooler ocean. The surface sea breeze is blowing inland from the cooler sea to the warmer land. The entire circulation is about 1 to 2 kilometers deep or so. So let's look at a typical sea breeze case, and let's focus on southern California. This is the city of Los Angeles, California. Well, actually, it's a topographic map, where red indicates relatively higher elevation. I've noted 2 places on this map, where you see L.A. campuses between the ocean and downtown, a few miles from the coast, and the San Fernando Valley, SFV, which is tucked around and behind the Santa Monica Mountains. Now the colored field is surface temperature, but we don't really measure temperature at the surface. It's actually measured 2 meters or 5.5 feet above the ground. Green colors indicate cool temperatures, and when we see red, they'll be very hot.

I'm also showing you the surface winds. We measure surface winds at 10 meters, or 33 feet, above the ground. This is a plot for 12:00Z on August 1. In meteorology, we mark time by London time. And 12:00Z is noon in London, so it's 5:00 am Pacific Daylight time. At this time, it's cool everywhere, and we see the winds are pretty weak. By 19:00Z, noon in Los Angeles, the land has warmed up a lot. There's been relatively little temperature rise over the sea, a consequence of seawater's very large thermal inertia. The winds have started blowing from sea to land. By 0:00Z August 2, it's midnight and a new day in London, but it's still August 1 in Los Angeles and 5:00 pm.

The lengths of these wind vectors indicate the wind speed, and you can see now the winds are much stronger. They're headed inland. It's very hot inland, particularly in the San Fernando Valley. Meanwhile, at UCLA, we're a lot cooler because the sea air is helping to keep us cool. After sunset, the land cools off very quickly, another consequence of the low thermal inertia of land. The temperature difference disappears, so the winds die down. It's still warmest in the San Fernando Valley. This is the hot part of town. The reason for this is because the sea breeze reaches there last, and is the weakest by the time it gets there.

If at night the land surface becomes colder, we will develop a land breeze. The circulation will reverse. Like all winds in meteorology, winds are named by where they come from. The land breeze blows from land to sea, just as the sea breeze blew from sea to land. So let's recap. Here are our sea and land breeze circulations. Temperature differences make pressure differences, and pressure differences drive winds. We have warm air rising over the heated land, and sinking over the cool sea, and the sea breeze circulation. It's opposite at night in terms of direction, but it's still warm air rising over the relatively warmer place, and sinking over the relatively cooler place. We call this "thermally direct," the better, more natural way for circulations to develop in response to temperature differences.

You may be wondering how these breezes actually work to decrease the temperature differences. There are several ways, the first of which leads us to the concept of wind chill. Wind chill is accelerated heat exchange, owing to wind. Picture a cold day. You feel much colder when it is windy than when it is calm, but if you take the air and you blow it around, particularly blow it at yourself harder, that doesn't make the air's temperature change, so why do you feel colder? Well, let's consider a familiar example: hot soup in a cold room. The soup is losing heat energy to the room air via conduction. What this is doing is warming up the air that is in direct contact with the soup, but air is a lousy conductor and it doesn't carry the heat energy that it acquired from the soup vertically away very efficiently. As a result, the temperature difference between the soup and the air actually decreases as a function of time.

This decreases the heat loss because the heat loss is proportional to the temperature difference. The air is actually acting as an insulating blanket, helping to keep your soup warm. Well, you don't want warm soup. You want to be able to eat it, and it's too hot, so what do you do? You blow that insulating blanket of air away, replacing the warmed air, the air that's been warmed by the soup, with relatively cooler air. What you've done is you have subjected the soup to wind chill. The wind chill effect depends on air being a poor conductor and a good insulator. In the case of the sea breeze, the cool sea air helps hasten the heat loss from the hot land as it blows across.

There are at least 2 other mechanisms for helping to reduce the land/sea temperature difference operating in this particular case. First, mixing of land and sea air will moderate the temperatures of both, and reducing the temperature difference from land to sea. The second actually represents the easiest and most important of all ways of changing the temperature of air. First, we need to think about a few things about air. We need to recall that air is very compressible and that compressed air gets warmer. Let's also recall that pressure decreases upwards, which means it increases downwards. This means if I take air and I force it to descend, it is encountering higher pressure. This is mechanically compressing the air. As a result, descending air warms. Air warms in descent. In fact, the easiest way to change the temperature of air is not to subject it to heating or cooling, but instead change its elevation.

Descending air warms at a very rapid rate, 10°C per kilometer, almost 30°F per mile. Here's an example. Suppose a mile above my head, the temperature is 70°F. I grab a piece of this air and I bring it down 1 mile to where I am. By the time I get it to my elevation, the temperature of that air is 100°F. This is temperature change due to volume change alone. There has been no heat transfer. We call this the "dry adiabatic" process. "Dry" means no role for moisture. "Adiabatic" means impassible or impossible. What is impassible here? This is temperature change without addition or removal of heat. This reflects air's compressibility and 3 things. First, you can change the temperature of air by changing its volume. Second, you can change volume by changing pressure. Third, you can change pressure by changing altitude because pressure varies with height.

Just as descending air warms because increasing pressure causes its volume to contract, so too does ascending air cool at the same rapid dry adiabatic rate of 30°F per mile, 10°C per kilometer. Rising air encounters increasingly lower pressure, permitting volume expansion. So the important point here is this: In the absence of a heat source or sink, rising air cools and sinking air warms up. So what we see is the vertical motions in the sea breeze circulation are actually working to decrease the temperature difference that drove the circulation to begin with. Air sinking over the sea is warmed by compression. The rising air over the warm land is cooled by expansion. The sea air is still cool, the land air is still warm, but the temperature difference is less than it would have been without these dry adiabatic effects. This is why we call the circulation "thermally direct."

This is why the circulation is thermally direct. The vertical air motions are contributing to reducing the temperature differences. So why is it more sensible for the surface wind to blow from cool to warm than from warm to cool? Picture the opposite circulation, where the air is sinking over the hot land, compressing and becoming hotter still. Picture the cool air rising over the cool sea, actually expanding and becoming cooler. We call that "thermally indirect."

Can air move in such a thermally indirect way? Yes, if it's forced, but that's not the natural way of doing things.

The sea breeze and the land breeze are both cool winds, but the winds don't always stay cool. My next subject is the Santa Ana winds. These have been described as "Those hot, dry winds that come down through mountain passes, curl your hair, make your nerves jump, and your skin itch." This quote came from a story by Raymond Chandler, who wrote stories based in the 1930s, 1940s, and 1950s in Los Angeles. Most of his stories involved his iconic private detective Philip Marlowe, but the Santa Ana winds were also a very common character in his stories. In fact, one of the nicknames for the Santa Ana winds is the "red wind," in addition to the "devil wind," and this was from a Chandler story called "Red Wind."

Santa Ana winds are the infamous winds of southern California. They often blow dry, hot, and fast. They're said to cause bad hair and worse moods. I'm not sure about that, but I do know that they dry out vegetation and they start and fan flames. On January 22, 2007, the Santa Ana winds were blowing, luckily, not too strongly. Luckily for me, that is. I was driving west towards home and there was a glow on the horizon. It was too late to be sunset. A glow like that on Santa Ana nights means fire. As I got closer, I realized that the glow was from my town. Closer still, I realized it was my neighborhood. It was even closer than that. The hills behind my house were on fire. After the cars were packed up with important papers and photos, I went back to take a few pictures of my own. The flames were as close as 150 yards, 137 meters, from my back fence. I could feel the spray from the water-dropping helicopters being carried by those dry winds. With a long lens, I could see the firemen, who stood between the fire and our neighborhood. In one of these pictures, you see a fireman setting a backfire to deny the advancing flames more fuel. Sometimes you have to fight fire with fire.

The Santa Ana winds are hot, dry winds that blow from the desert, but they blow when the desert is cool, so how do they get hot? Most Santa Ana events start with cold, dense air, spilling down across the Great Basin of Nevada and Utah. This cold air is trapped on the west by the Sierra Nevada Mountains, which are a tall and formidable obstacle. The southward progress towards Los Angeles is partially stopped by a ring of mountains surrounding the Los Angeles basin.

Here's another way of looking at the topography of southern California. I'm showing you the locations of Los Angeles and the Mojave Desert to the northeast, and the Great Basin of Nevada. We see the ring of mountains, which surround our basin.

Now, the colored field is the density of the air at the 850-millibar level. The 850 level is about 1.5 kilometers, or 1 mile, above sea level. A red color means less dense, blue is denser. Here are the 850-millibar winds. At this time, the air over southern California has a low density and Los Angeles has a sea breeze. But all that changes pretty quickly. Watch as the tide of cold, dense air sweeps down from the north on the east side of the Sierra. The Sierra is preventing those winds from spreading westward. This is density at 850 millibars, but you can consider that as temperature. Recall the ideal gas law. At the same pressure, warm air is less dense than cold air, so the blue color actually not only means denser, but also colder.

As the cold air reaches the mountains surrounding Los Angeles, it's stopped, sort of. Some is sneaking through the passes and canyons, making the winds fast. Think of a garden hose. The flow is coming out of your garden hose. You can use your finger to create a restriction, and that makes the velocity actually faster, doesn't it? This makes the flow faster, but it does not make the winds hot. The situation is a lot of the air is actually passing over the mountain, and then diving down into the L.A. basin. The air has gone downslope. It has experienced compression and it has warmed 30°F per mile of descent, 10°C per kilometer.

It started off cool, but then it got hot. So Santa Ana winds start off cool, but they don't stay that way. This is a type of wind we call "katabatic wind." In Greek, this means to flow downhill. It flows down, owing to its initially larger density, initially greater weight. Now, descending air warms up at 30°F per mile. That's 10°C per kilometer of descent. On ascent, dry air cools at that same rate. This is a huge number, and it's hugely important to meteorology, but I haven't told you yet why it's 10°C per kilometer, where it came from.

This is the ratio of 2 important numbers, g and cp where g is the acceleration of gravity. This tells us how quickly an object in freefall gains velocity on descent. It's 32 feet per second per second, 10 meters per second per second, or 10 meters per second squared. And cp is a measure of thermal inertia. The larger it is, the slower an object will heat up or cool down. For air in the metric system, it's 1004 joules per kilogram. A joule is a unit of energy. Remember, air warms on descent because the pressure squeezing on the air increases. Why does gravity matter? Because without it, there would be no pressure at all, and air would have no weight. So the larger gravity is, the faster pressure changes with elevation.

How do we increase gravity? Well, make the Earth more massive, more like Jupiter. Expansion and compression can change temperature, but cp controls how quickly. So cp is the specific heat of the air at constant pressure, and as I said, it measures thermal inertia, so the 30°F per mile, or 10°C per kilometer, rate, which is so important in meteorology, reflects the Earth's mass—that determines gravity—and the Earth's composition. If we change it, we change cp.

Let's summarize this lecture. We started with a statement: Temperature differences make pressure differences, and pressure differences drive winds. The driving force, pressure, is gravity force per unit area, reflecting the fact that air has weight. Air pressure decreases with height because there's less air to push down as we ascend. Pressure increases downward towards the ground for the same reason. There's more air above us, pressing down. The not-fully-realized goal of winds is to remove the temperature imbalances that created the pressure differences to begin with.

We saw a simple example, where unequal heating led to a circulation, the sea breeze, owing to the relationship between temperature and pressure. Layers of air became thicker upon being heated. This created horizontal pressure differences that started the air in motion. The sea breeze and its nighttime counterpart, the land breeze, are examples of where the surface winds blow from cold towards warm, but it's part of a circulation where, in other places, the winds are blowing in the opposite direction. In some places, air is rising, and in other places, air is sinking. But even winds that start out cold may not stay that way because air is very compressible. Forcing air to descend makes its temperature rise, since the air pressure pushing on the air is increased. This is why the Santa Ana winds can be so hot after they are impelled downslope. Those are hot dry winds that blow from the desert, but they do so when the desert is cold.

So let's look ahead. Santa Ana winds represent a fire hazard, in large part because they're dry. Why are they dry? The air starts off in the desert, where it is not only cool, but it may be relatively moist as well. In the next lecture, we will see that temperature is a powerful control on humidity. Bringing air downslope not only raises the temperature of air, but it also decreases its relative humidity. Now for the flip side: Forcing air to rise increases its humidity, as the air expands and the pressure decreases. So we can force air to become saturated on ascent. We can force water vapor to condense.

Why does rising air lead to clouds? Once clouds form, does this change the behavior of how air changes on ascent? We'll start looking at moisture in the next lecture.

Lecture Seven
An Introduction to Atmospheric Moisture

Scope:

In the last lecture, we examined sea breeze and land breeze circulations. We also learned that the Santa Anas are hot because of downslope motion and fast because they're channeled through the mountains. In this lecture, we'll look at atmospheric moisture to find out why the Santa Ana winds are dry. As we go through the lecture, keep in mind these questions: Why are we susceptible to static electric shocks when we walk around in stocking feet? And why are clouds the likely result of forcing air to ascend?

Outline

I. Water vapor is a great place to start our exploration of humidity.

   A. The ability of air to hold water vapor is a strong function of temperature. Warm air can hold more water vapor than cold air, which explains why most vapor is located close to the ground.

   B. A sample of air has 2 moisture properties—vapor capacity (VC), which tells us how much water vapor air can hold at a maximum, and vapor supply (VS), which tells us how much water vapor exists in the sample.

   C. A sample of air with 10 grams of water vapor per kilogram of dry air would have a vapor supply of 10 grams per kilogram (10 g/kg).

   D. Relative humidity, or RH, is the ratio of vapor supply and capacity expressed as a percentage.

   E. Vapor supply is a measure of absolute humidity, which is primarily a function of temperature. For each increase of 10°C, vapor capacity approximately doubles.

   F. It doesn't take much water vapor to saturate cold air, and if extra water vapor is present, it forms liquid or ice.

II. The Santa Ana winds become hot due to the dry adiabatic process, which involves temperature change from volume contraction rather than heat exchange.

    A. A sample of descending dry air will heat at a rate of 10°C per kilometer. Moist air, that is, air with some vapor supply, also compresses dry adiabatically, as long as the air is cloud-free.

    B. As the moist air moves downslope, its vapor capacity increases dramatically, due to compression warming and the exponential relationship between vapor capacity and temperature.

    C. As a Santa Ana wind moves downslope, both its temperature and its vapor capacity increase, but the vapor supply remains the same; thus, the relative humidity drops.

    D. Indoor relative humidity can become uncomfortably low in the winter, resulting in static electric shocks. Under higher humidity conditions, electric charge can't build up because objects are coated with a very thin layer of surface water.

    E. Even if the water content of air doesn't change, relative humidity will change with temperature. Relative humidity will be highest when the temperature is lowest.

III. Let's make a cloud by doing our Santa Ana example in reverse.

    A. Suppose air starts at the foot of a 2-mile mountain under these conditions: temperature, 110°F; vapor capacity, more than 50 g/kg; and vapor supply, 8 g/kg (relative humidity under 16%).

    B. Forcing that air to ascend the mountain decreases the temperature and the vapor capacity. With just a 2-mile lift, we can bring the air to saturation through expansion cooling.

IV. Why does vapor capacity increase with temperature? The answer involves the temperature of water molecules and the pressure they exert.

    A. The 3 phases of any substance are vapor, solid, and liquid. We'll consider vapor and liquid at temperatures above freezing.

    B. Water molecules are always in collision. The lower the temperature, the less likely it is that 2 vapor molecules can avoid bonding when they collide, so the drops they form are more likely to be long-lived.

C. The higher the temperature, the more likely it is that a water molecule can escape from its liquid prison and become vapor.

D. Water vapor molecules in the vicinity of a plain surface of liquid water, such as a lake or a pond, exert a pressure on the liquid water surface, known as vapor pressure.

E. Vapor molecules in motion may become incorporated into the liquid water surface (condensing), while water molecules in the surface seek to return to the atmosphere as vapor (evaporating). When the rates of condensation and evaporation are the same, the vapor liquid system is in equilibrium, also called saturation.

F. Actual vapor pressure is the vapor supply. Saturation vapor pressure is the vapor capacity. These concepts are expressed as mass or quantity, rather than pressure.

G. As the temperature gets higher, more vapor molecules are needed to balance the escapees. That balance is saturation, and the saturation vapor pressure and vapor capacity increase quickly with temperature.

H. In a nutshell, the ability of air to hold water vapor is really the ability of vapor to avoid condensing and succeed in evaporating.

V. How does the condensation process start?

A. Vapor molecules require a surface to condense on, such as grit suspended in midair. Such surfaces are called condensation nuclei.

B. Condensation particles start off as small cloud droplets that fall very slowly relative to still air. But collisions amongst these droplets can result in larger, faster-falling drops, which may grow by accreting the smaller particles.

C. To become ice, liquid water needs an ice nucleus, but ice nuclei are not as common as we might expect. Thus, liquid water drops, called supercool liquid, can persist in subfreezing temperatures.

D. The Bergeron or ice crystal process for precipitation production is based on the fact that ice crystals will grow rapidly at the expense of supercool drops once they manage to form.

E. The saturation vapor pressure over ice is smaller than that over water, allowing ice to survive in conditions that would cause liquid water to disappear. In a nutshell, ice outcompetes liquid in subfreezing temperatures, but it starts with a disadvantage—a lack of ice nuclei to kickstart the process. Cloud seeding represents an attempt to supply the missing ice nuclei.

Suggested Reading:

Vonnegut, *Cat's Cradle*.

Questions to Consider:

1. True or false: Water vapor can condense to liquid in sub-saturated air. Why?

2. In the premier episode of the TV series *Star Trek:Voyager*, a powerful entity said he performed an experiment on a populated planet that inadvertently rendered it a desert. His experiment removed all particles from the planet's atmosphere that permitted water vapor to condense upon them, permanently ending the possibility of rainfall on the planet. Does this scenario make sense?

Lecture Seven—Transcript
An Introduction to Atmospheric Moisture

Welcome back to Meteorology. In the last lecture, we examined the sea breeze and land breeze circulations. These are local circulations, driven by temperature differences. The low thermal inertial land heats up quickly during the day, despite actually absorbing less solar radiation than its neighbor, the sea. The heated air gets thicker and deeper, creating a horizontal pressure difference directed from land to sea above the ground. Air above the ground starts flowing seaward, the first link in a circulation loop that ultimately ushers cool sea air inland along the surface. At night, the sea breeze switches to a land breeze, as the land cools quickly, and again, that's low thermal inertia.

So both the daytime sea breeze and the night breeze are cool winds, but we saw that the land breeze can be hot if forced to flow downslope. That was the ultimate land breeze, the Santa Ana. The Santa Anas are hot, owing to downslope motion, which cause compression. The winds could actually start out quite cool, but they warm up quickly. The winds can speed up too as they're channeled through passes and canyons. But neither of those things tells us why the Santa Ana winds are so dry. This lecture is our first look at atmospheric moisture.

Some questions to think about during this lecture: When are we susceptible to getting static electric shocks when we shuffle around in stocking feet, and why? Why can ice appear on airplane wings? Shouldn't ice just bounce off? Why are clouds the likely result of forcing air to ascend? Let's get started.

I've already made this important statement: The ability of air to hold water vapor is a very strong function of temperature. That means that warm air can hold much more water vapor than cold air. That's one reason why water vapor is not well mixed through the atmosphere. Most of the vapor is located pretty close to the ground because the air further aloft is too cold to hold it. Eventually, I will be qualifying this statement, but it's a great place to start our exploration of humidity.

A sample of air will have 2 moisture properties—vapor capacity (VC), which tells us how much water vapor air can hold at a maximum, and that's primarily a function of temperature, and vapor supply (VS), which tells us how much water vapor actually exists in the sample. We can express these 2 properties as mixing ratios. A mixing ratio is a mass of water vapor divided by a mass of dry air, excluding the vapor. It's like a recipe. For example, in my backyard, I have a hummingbird feeder, and in it I put a mixing ratio of 1 part sugar to 4 parts water. For convenience, since water vapor content is so small, we express the vapor mass in grams and the dry air mass in kilograms. Suppose a sample of air had 10 grams of water vapor per every 1 kilogram of dry air, its vapor supply would be 10 grams per kilogram. It's just understood by us that the grams refers to water vapor and the kilograms to dry air.

Is 10 grams per kilogram of vapor supply a lot, or a little? Well, actually, it depends on what the sample's vapor capacity is. Relative humidity, or RH, is the ratio of vapor supply and capacity expressed as a percentage. Let's take our sample of air with a vapor supply of 10 grams per kilogram. If the vapor capacity of that air is also 10 grams per kilogram, the relative humidity is 100%, and we're at saturation. That air can hold no more water vapor. If the vapor capacity of that air were 20 grams per kilogram, the relative humidity would only be 50%, and that air would not feel especially damp. If the vapor capacity were 30 grams per kilogram, as it might be on a warm summer day, the relative humidity is now only 33%, and that might feel uncomfortably dry to you.

Next, we should appreciate that vapor supply is a measure of absolute humidity, the actual vapor content of the air. The same absolute humidity of 10 grams per kilogram might feel damp or arid to you, depending on the air's vapor capacity, and that's primarily a function of temperature. The bottom line that we'll be seeing again and again is this: You have a good feel for relative humidity, but you're a very poor judge of vapor supply.

Here are some numbers, some vapor capacity values for air at sea level. The numbers themselves aren't important, but the pattern is. I'm going to start at −20°C, and I'm going to increase by intervals of 10°C, or 18°F. Minus 20°C is −4°F, and that air can hold 3-quarters of a gram per kilogram. But at −10°C, or 14°F, that air can hold 2 grams per kilogram at a max. I've gone from 3-quarters of a gram per kilogram to 2 grams per kilogram. I've more than doubled. At 0°C, or 32°F, that air can hold 4 grams per kilogram. We've doubled again. At 10°C, or 50°F, 8 grams per kilogram. At 20°C, or 68°F, 15 grams per kilogram, a little less than double. At 30°C, 86°F, we're at 28 grams per kilogram. At 40°C, or 104°F, that air can hold 50 grams per kilogram at a maximum, an increase of 66 times over the air at −4°F.

These values demonstrate that vapor capacity is a very strong function of temperature. It's an exponential function. For each increase of 10°C, or 18°F, vapor capacity approximately doubles. If you increase the temperature a little, its vapor-holding capacity increases a lot. Vapor capacity is also influenced by pressure, to some degree, but we're effectively ignoring that for now. Vapor capacity is also called "saturation mixing ratio" since it represents the mixing ratio or vapor supply that would be present in air if it were saturated. At saturation, air is holding all the water vapor it can. Our vapor capacity numbers clearly show that it doesn't take much water vapor to saturate cold air, and if extra water vapor is present, it has to condense, forming liquid or ice, depending on conditions.

It's easy to make clouds or fogs out of cold air, but that also means it's very hard to get much precipitation out of very cold air. It may be saturated. The relative humidity may be 100%, but the vapor supply values are so small that precipitation has to be small. It's like getting blood out of a turnip. In contrast, hot air is thirsty air. The vapor capacity of hot air is enormous, and it's very hard to saturate. Picture a thimble versus a swimming pool. That's a little bit of an extreme example, but we don't find hot saturated air very often, unless we're in a sauna. By the way, mixing ratio tells us what fraction of air water vapor represents. Recall dry air is 78% nitrogen, and 21% oxygen, 1% argon and so on. A vapor supply of 10 grams per kilogram means air is 10 parts water vapor for every 1000 parts dry air, so the water vapor content is actually just a little bit under 1%.

Let's do a non-intuitive example. Let's compare a foggy winter day in Minnesota, relative humidity 100%, with a hot, very dry, early summer day in Death Valley. Chances are, there's more water vapor in the Death Valley air than in that Minnesota fog. In Minnesota, our temperature might be −10°C, or 14°F. The vapor capacity is 2 grams per kilogram. I said we're saturated, so the vapor supply is 2 grams per kilogram also. We have a fog. In Death Valley, suppose the temperature is 40°C, or 104°F, so that's probably a little before lunchtime. The vapor capacity of that air is 40 grams per kilogram, and if the relative humidity is 5%, that's oppressively dry. This means the vapor supply is 2.5 grams per kilogram, more moisture in Death Valley than in Minnesota. Again, you're a poor judge of absolute humidity.

In the last lecture, we discussed the hot, dry Santa Ana winds of southern California. These were the hot, dry winds that blow out of the desert when the desert is cold. They became hot because they traveled downslope and experienced compression warming. We call this the "dry adiabatic process," involving temperature change due to volume contraction, rather than heat exchange. A sample of descending dry air will heat up at a rapid rate, 10°C per kilometer, or almost 30°F per mile. Moist air, by which I mean air with some vapor supply, also compresses dry adiabatically, as long as the air is cloud-free. In fact, a better term for the dry adiabatic process would be "subsaturated adiabatic," but I'm going to stay with the time-honored term. As the moist air moves downslope, its vapor supply may not change, but its vapor capacity increases dramatically, due to the compression warming and the exponential relationship between vapor capacity and temperature.

Let's do an example. Santa Ana winds originating over in Nevada might have a temperature of 50°F, so that's a vapor capacity of 8 grams per kilogram. Let's say the relative humidity is 100%. That's an exaggeration, but that would mean the vapor supply is 8 grams per kilogram as well. Now, let's move that air downslope 2 miles into Los Angeles. The air's temperature increases by 60°F, becoming 110°F. Its new vapor capacity is more than 50 grams per kilogram, but the vapor supply is still only 8 grams per kilogram. So instead of being saturated as we started out, the relative humidity is now under 16%.

We had a drop of relative humidity from a very moist 100% to bone-dry 16%, just by moving the air downslope. So Santa Ana winds can start off cool and moist in the high-elevation Great Basin, but they have very low relative humidity by the time they reach sea level. In fact, relative humidity down to 5% is not uncommon when the winds are howling. This dry air is thirsty and it steals moisture from wherever it finds it, like vegetation, which is one reason why low relative humidity is so conducive to fire.

Let's do another example. If you live in a cold climate, you undoubtedly know that indoor relative humidity can become uncomfortably low in the winter. Let's go back to our foggy winter day in Minnesota. The temperature was 14°F, relative humidity at 100%, vapor supply and capacity at 2 grams per kilogram. Let's bring this air indoors and heat it up to a reasonable room temperature of 68°F, or 20°C, but without changing its vapor supply. Its vapor supply is still 2 grams per kilogram, but the heated air's new vapor capacity is 15 grams per kilogram, so our new relative humidity is 2 divided by 15 or 13%, shockingly low, literally. When the relative humidity gets down to values below 20%, static electric shocks become a common problem. If you shuffle your feet on a carpet and then touch a piece of metal, an electric shock can be created. Friction can caused an electric charge to be created on your body.

Under higher humidity conditions, this charge cannot build up. That's because objects become coated with a very thin layer of surface water, even when the relative humidity is way below 100%. Water isn't just a good conductor of heat—it conducts electricity very well, and as a result, the charge cannot hold and it cannot build up. But when the air is relatively dry, the water coat is absent, and the charge can build up to be discharged when you touch another good conductor, like metal. So the solution to your little indoor problem is to humidify the air, adding water vapor to increase the relative humidity.

Nature has a big outdoor problem with static electricity build-up because air is a lousy conductor of electricity. Nature's discharges of static electricity tend to be dramatic and deadly, in the form of lightning. We'll examine how and why lightning forms later in the course. We defined adiabatic as temperature change due to volume and pressure changes. No heat exchange was involved. If I heat air in a furnace, that's an example of a diabatic process, temperature change by heat exchange, so diabatic is the opposite of adiabatic. The source of the temperature change doesn't matter. If you increase the temperature a little, the relative humidity decreases a lot.

So my examples so far have shown that even if water content of air doesn't change, relative humidity will change with temperature. Did you notice that it's an inverse relationship? All our examples so far have involved warming the air by heat exchange or compression, so it's time to do the reverse. Suppose the vapor supply does not change during the day. What time of day should the relative humidity be highest? Well, it'd be highest when the temperature is lowest, and on a typical day, that'll be around sunrise because the ground has been cooling all night. In this way, we can increase relative humidity by lowering the air temperature, and with enough cooling, we can potentially raise the relative humidity all the way to 100%. As we've seen, a great way to change temperature is to change its elevation.

Lifting air causes it to expand. That was adiabatic expansion cooling. So as long as the air has a little bit of moisture, even just a smidgeon of moisture, we can lift that air to saturation, if we can have the means to lift it.

Let's make a cloud. Let's do our Santa Ana example in reverse. Suppose air starts at a foot of a 2-mile tall mountain on a hot and relatively dry day. Let's say the temperature is 110°F, 42°C. The vapor capacity of that air is more than 50 grams per kilogram, and let's say the vapor supply is only 8 grams, so the relative humidity is under 16%. Now we force that air to ascend to a mountaintop 2 miles high. It cools 60°F on the way up, all the way down to 50°F, or 10°C. Its vapor capacity has decreased from a formerly enormous value down to a humble 8 grams per kilogram. That's the same as the vapor supply, so our relative humidity is now 100%. We started very hot and relatively very dry, and with just a little bit of lifting, and 2 miles is not that much, we've brought this air to saturation through expansion cooling. Any further lifting would cool the air more, causing the vapor capacity to become smaller than its vapor supply.

Now, with a vapor excess in that air, that air parcel that started off brutally dry, has now become a cloud. A cloud has been born. Take a look outside. There may be a deep cloud right above your head. The air at that cloud's apex likely originated near your feet, so when you look at a cloud, you should be thinking air has risen there.

Why does vapor capacity increase with temperature? Why does warm air hold more water vapor than cold air? Actually, it doesn't have anything directly to do with the temperature of nitrogen and oxygen or argon, and everything to do with the temperature of the water molecules themselves, and the pressure they exert. Let's investigate this.

The 3 phases of any substance are vapor, solid, and liquid. For simplicity, let's just consider vapor and liquid at temperatures above freezing. One thing we need to remember is water substances are always moving between these 2 phases, always going back and forth. Vapor molecules are becoming liquid. They're condensing. Liquid drops are becoming vapor. They're evaporating. The microscopic activity of a water molecule is a function of its temperature, and increases quickly with its temperature as well. Water molecules are always in collision. The lower the temperature is, the less likely 2 vapor molecules can avoid bonding upon collision, so the drops that they form are more likely to be long-lived. The higher the temperature, on the other hand, the more likely that a water molecule can escape from its liquid prison and become vapor, so there will be very few surviving drops.

Let's examine this more closely, but first in the context of pressure rather than temperature. Consider water vapor molecules in the vicinity of a plain surface of liquid water, like a lake or a pond. These vapor molecules are exerting a pressure, force per unit area on the liquid water surface, and we call this "vapor pressure." It's a small part of the total atmospheric pressure exerted on the liquid because vapor represents only a very small part of the air's mass. But if we increase the number of vapor molecules in the air, if we increase its amount, we would increase the vapor pressure. The vapor pressure is a function of the water vapor mass.

Now let's turn to temperature. Temperature is measuring the microscopic energy and vibration of translation of water molecules in the air and in the liquid. The vapor molecules in motion may chance to collide with the liquid water surface. If and when they do, there's always a chance they'll become incorporated into the liquid water prison. They will have condensed. Meanwhile, water molecules in the liquid water prison seek to escape, to break the surly bonds with their neighbors and return to the atmosphere as vapor. If and when they succeed, they will have evaporated. When the rates of condensation and evaporation are the same, the vapor liquid system is in equilibrium. There's vapor becoming liquid, there's liquid becoming vapor, but there is no net gain or loss of either form of water. This equilibrium is called "saturation."

At saturation, there's just enough vapor molecules around to replace the water that escapes from the liquid. This vapor mass exerts a pressure, and the pressure is called the "saturation vapor pressure." It is simply the vapor pressure exerted at the condition of saturation. What happens if there are too few vapor molecules around to replace those that managed to escape from the liquid water prison? This situation is subsaturation. The liquid water surface is losing more vapor molecules than it's gaining, so it is evaporating. There's a net transfer of water mass from liquid to vapor. The shortage of vapor mass also means that the vapor pressure is smaller than it needs to be. This means that the actual vapor pressure is less than the saturation vapor pressure. Actual vapor pressure represents supply.

Saturation vapor pressure represents demand, need, or capacity, and indeed, for our purposes, actual vapor pressure is the vapor supply, and saturation vapor pressure is the vapor capacity.

Vapor supply and vapor capacity are moisture concepts expressed as mass, or quantity, rather than the consequence of mass, which is pressure. So how does temperature fit in? As the temperature gets higher, more vapor molecules would be needed to be present in order to balance the escapees. That balance is saturation, and the saturation vapor pressure and vapor capacity increase with temperature. Since the microscopic energy increases quickly with temperature, so does saturation vapor pressure and vapor capacity. They're exponential functions. The relevant temperature here is not the temperature of the nitrogen and the temperature of the oxygen per se, but because of collisions and heat conduction amongst the constituents of air, it's likely that the water molecules have the same temperature as their neighboring constituents. So in a nutshell, the ability of air to hold water vapor is really the ability of vapor to avoid condensing. It's the ability of water vapor to succeed in evaporating.

Now, we're not just measuring the temperature of water vapor alone, so we will persist in our convenient untruth about air holding vapor. How does the condensation process start? It turns out that vapor molecules require a surface to condense on. Large existing liquid water surfaces, like the lake we've been considering, definitely work, but you don't find those everywhere. Other useful surfaces are bridges, picnic tables, your skin, blades of grass, and your glasses if you have any. But what about in the middle of the air? What kind of surfaces exists there? Well, you'd be surprised.

There's grit suspended in the air, grit like sand particles, dust, soot, salt crystals, even in what we call clean air, and all of these surfaces in midair can serve to support cloud droplet formation when the conditions are ripe, and we call them "condensation nuclei." Look inside each cloud droplet. There's at least one piece of dust, salt, soot, or whatever, and raindrops are made of many cloud droplets, and so contain many pieces of grit. I'm sorry to be the first person to have to tell you that pure rainwater isn't pure. Some surfaces make very good hosts for condensing water molecules, and we call them "hygroscopic nuclei." "Hygro" is a combining form, meaning water. "Scopic" means to see or seek. These are water-seeking nuclei.

Particles formed by sulfuric and nitric acid pollutants, soot particles, and other things like that help make haze in polluted cities because they attract water that might not otherwise condense, they are hygroscopic nuclei. But the best condensation nuclei, the most hygroscopic nuclei of all, is salt. Have you ever noticed that it is often hazy along the seashore, even when the air is not polluted? We'll talk about the optical phenomenon of haze, when it's blue, when it's white, and why, in a future lecture. But you're likely seeing small condensation particles. Salt is a great condensation nucleus. Indeed, water vapor will start condensing on salt particles at relative humidity as low as 75%. That's what is making salt clump in your salt shaker. Did you ever go to a restaurant and look in their salt shakers? A trick they use in restaurants is to put rice grains in the salt shaker because rice grains very readily absorb water. They're even better at absorbing water than salt is, so that helps keep the salt from sticking.

On the other hand, hydrophobic nuclei repel water. "Hydro" is another prefix for water. "Phobic" means to fear. Now, this is not rabies, a disease also called "hydrophobia," but water does seem to have a rabid fear of these hydrophobic nuclei. Examples are oils, waxes, and Teflon. Look on your new, nicely waxed car. You can see water beading. Look closely. That water drop is practically on its virtual tiptoes, trying to get away from that waxy surface. Now, hygroscopic or hydrophobic, there is no shortage of condensation nuclei in the atmosphere, even in clean air.

Condensation particles start off as small cloud droplets, and these small particles fall only very, very slowly, relative to still air. But collisions amongst these droplets can result in larger, faster-falling drops, and larger drops falling through a group of smaller particles may grow by accreting these droplets. They coalesce. This is the collision coalescence or warm rain process. Warm rain means it does not involve ice. We'll consider frozen water next.

We all know that 0°C and 32°F is the freezing point of water, but I should say the so-called freezing point. Liquid water may not, and often does not, freeze right away. The reason is at subfreezing temperatures, tiny ice crystals try to form, but they're fragile and often rapidly disintegrate. In fact, pure undisturbed water can resist freezing all the way down to a temperature of about −40°C, or −40°F. At that temperature, homogeneous nucleation or spontaneous freezing occurs. Water droplets literally and finally freeze on themselves. But at higher temperatures, water needs a helper particle to freeze, what we call an "ice nucleus," and water is much more selective with regard to ice nuclei than it is with condensation nuclei. In fact, the best ice nuclei are particles that are structured like ice, such as ice itself, but until ice itself is abundant, ice has a hard time freezing. This is a real catch-22.

As a result, ice nuclei are not as common as one might expect, especially in the region of the cloud between temperatures of 0°C and −15°C, or 32°F and 5°F. The upshot is this—due to a lack of ice nuclei and the selectivity and slowness of the freezing process, liquid water drops can persist in subfreezing temperatures, and we call this "supercool liquid." This is a significant aviation hazard because supercool liquid can freeze on contact with aircraft wings and other parts, impairing its flightworthiness. In this situation, the airplane supplies the missing ice nucleus.

The Bergeron, or ice crystal, process for precipitation production is based on the fact that ice crystals will grow rapidly at the expense of supercool drops, once they manage to form. But recall that ice saturation represents the equilibrium state, where the gain and the loss of water is the same. In that case, there are enough vapor molecules to compensate for the escapees from the liquid. In other words, the actual vapor mass and pressure equal the saturation vapor mass and pressure. Subsaturation means there aren't enough vapor molecules around to compensate for those escaping from the liquid, but there's a difference between liquid water and ice. It's easier for water molecules to escape from prison. The bonds amongst the ice molecules are much stronger and much harder to break, so ice doesn't need as many vapor molecules hanging around to compensate for the escapees, at least when the temperature is below freezing.

In other words, the saturation vapor pressure over ice is smaller than that over water, and it also means that ice can survive in conditions that would cause liquid water to disappear. In a nutshell, ice outcompetes liquid in subfreezing temperatures. But remember, ice starts with a serious disadvantage—a lack of ice nuclei to kickstart the process, particularly in the dead zone between 5°F and 32°F, where ice nuclei are rare. That's the idea behind cloud seeding. This represents an attempt to supply the missing ice nuclei.

The first attempts at cloud seeding used dry ice, frozen carbon dioxide. Dry ice is extremely cold, −78°C, −104°F. This causes liquid to freeze spontaneously, to homogeneously nucleate. Subsequently, an atmospheric scientist named Bernard Vonnegut experimented with silver iodide as a cloud seeding agent. Silver iodide has a crystal structure resembling ice, and thus promotes the formation of more ice. Vonnegut's research inspired his younger brother Kurt Vonnegut to write *Cat's Cradle*, a story of the ultimate ice nucleus. Considerable work on cloud seeding continues, and it remains controversial.

So let's summarize. Warm air can hold much more water vapor than cold air. We can express this as a vapor capacity, VC, telling us how many grams of water vapor we can get into every kilogram of air. Actual vapor content, also in grams per kilogram, is vapor supply, VS, and relative humidity is the ratio of the 2. Actually, air does not hold vapor. That's a convenient fiction. In truth, as the temperature of vapor rises, it's more able to resist bonding with its brethren on collision. That doesn't change the fact that the vapor capacity of warm air is enormous. At 100°F air can contain 12 times more vapor than air at freezing, although it usually doesn't.

So vapor capacity is a very strong function of temperature, and air temperature is very easy to change by adding or subtracting heat, or by compressing or expanding air. For a fixed vapor content, relative humidity will rise as the temperature falls, so dew and fog are most likely to occur around sunrise, when the air is typically coolest. We can make clouds through lifting air because that causes expansion cooling.

Forcing air to descend, on the other hand, is a great way to make air hot, as well as relatively very dry. The air carried by the Santa Ana winds may start out cool and even relatively moist in Nevada, but by the time it drops to sea level, it's a completely different story. Both condensation and freezing processes need a nucleus, a surface to start on. Condensation nuclei are abundant. Ice nuclei are much more scarce, leading to the existence of supercool liquid in clouds, with temperatures below freezing. Liquid water is still present, owing to a lack of ice nuclei, until an airplane flies through, the mother of all ice nuclei. Cloud seeding is an attempt to augment the natural lack of these nuclei.

Let's look ahead. There's more to the story of precipitation, how it forms, how it grows, how it falls, and we'll return to this. Our next step, however, is to use our useful concepts of vapor supply and vapor capacity to understand the ways in which air can be brought to saturation. This will lead us to an understanding and appreciation of how clouds and storms form, when, where, and why.

Lecture Eight
Bringing Air to Saturation

Scope:

In the last lecture, we introduced the central issue of atmospheric moisture, and we saw how to saturate air through expansion cooling. In this lecture, we will see 2 more ways of bringing air to saturation, and we'll encounter 2 new kinds of temperature. Along the way, we'll consider a few questions: What is black ice, and when and where does it form? Why are ice cubes you make in your freezer so often cloudy in the center? And why does it take food longer to cook at higher elevations?

Outline

I. Let's review vapor supply and vapor capacity.

   A. Vapor supply (VS) is the amount of water vapor in air, expressed as grams of water vapor per kilogram of dry air.

   B. Vapor capacity (VC) represents the amount of water vapor air can hold at a maximum. The ability of air to hold water vapor is a strong function of temperature.

II. In addition to the dry adiabatic process, there are 2 other ways to saturate air, which introduces us to the dew point temperature and the wet bulb temperature.

   A. We can understand the dew point if we think about a cold can of soda "sweating" on a hot day.

   B. As we have seen, the regular or dry bulb temperature tells us the vapor capacity of air. The dew point temperature tells us the air's vapor supply.

   C. At saturation, vapor supply and capacity are the same, so the temperature and the dew point are also the same.

III. We can bring air to saturation without changing its vapor content or pressure by extracting heat. This is a diabatic process. When the temperature of air is decreased to its dew point, saturation is reached.

A. Suppose we have a subsaturated sample of air. If the temperature is higher than the dew point, the vapor supply is less than the vapor capacity. If we decrease the temperature of the air, the vapor capacity decreases and the relative humidity increases.

B. When the relative humidity reaches 100%, any further cooling will make the air supersaturated, and condensation will appear. The dew point approach to saturation—isobaric cooling—does not change the vapor supply.

C. On most nights after sundown, the temperature of air near the ground cools, decreasing the vapor capacity. If the air cools to its dew point, condensation will form, using the ground surface as condensation nuclei.

D. Dew or frost typically forms around sunrise and on objects with low thermal inertia, such as car windows and concrete bridges. Black ice is a particularly dangerous kind of ice that forms on roadways when subfreezing air is chilled to the dew or frost point.

E. Here's a forecasting rule: If the air mass at your location is not expected to change overnight, the minimum temperature will probably be no lower than the afternoon dew point. If the air is cooled overnight to the dew point, further cooling will be more difficult because there is resistance to cooling in the form of condensation warming.

IV. Evaporation is a cooling process.

A. When you climb out of a swimming pool, you feel cold because the liquid water on your skin is evaporating into subsaturated air. It takes energy—heat coming from you—to break the bonds that hold water molecules together into liquid water drops.

B. The water evaporates because the air is subsaturated, and the vapor supply is less than the vapor capacity.

C. As evaporation proceeds, the number of vapor molecules around your skin increases, and the vapor supply and pressure both increase. As the difference between the vapor supply and the vapor capacity gets smaller, evaporation and the cooling it produces slows down.

V. Saturation by evaporation cooling and the subsequent moistening is the wet bulb approach to saturation.

   A. We cannot bring a subsaturated sample of air to saturation through evaporation, because evaporating water takes energy from the air cooling it, and as the air cools, the vapor capacity decreases.

   B. As we evaporate water into the air sample, the water vapor content and the vapor supply increase, but the temperature and the vapor capacity decrease. At some point, the temperature and the dew point will meet at a new temperature, called the wet bulb.

VI. We've looked at 3 temperatures: actual, dew point, and wet bulb.

   A. The actual temperature tells us vapor capacity.

   B. The dew point tells us vapor supply and shows how much we can cool the air without changing the vapor supply before saturation is reached.

   C. The wet bulb temperature tells us how much we can cool the air by evaporation.

   D. At saturation, all 3 temperatures are equal.

VII. Boiling is actually rapid evaporation.

   A. A bubble that forms in a pot of boiling water is subject to pressure from both the atmosphere and the liquid water. The boiling point is the temperature at which saturation vapor pressure pushing out from the bubble matches the external pressure pushing back.

   B. The fact that pressure decreases with height explains why the boiling point decreases with elevation above sea level.

   C. This illustrates an important point: Phase changes occur at constant temperature, as well as constant pressure.

   D. The phase transition between solid and liquid water also involves latent heating and cooling. This is called the latent heat of fusion or melting, depending on the direction of the process.

VIII. One of the principal roles of clouds is to transport heat vertically, with latent heat release fueling potentially powerful updrafts. Whether clouds can accomplish this, however, depends on how stable the environment is.

Suggested Reading:

There is no suggested reading for this lecture.

Questions to Consider:

1. You are taking a shower with warm water, some of which is evaporating into the room air. You notice that before the room air becomes noticeably foggy, condensation has already begun forming on your bathroom mirror. Why?

2. Under what atmospheric and surface conditions are we likely to get frost on the ground?

Lecture Eight—Transcript
Bringing Air to Saturation

Welcome back to Meteorology. In the last lecture, we started exploring the crucial and central issue of atmospheric moisture. In particular, we appreciated that the ability of air to hold water vapor is a very strong function of temperature. We exploited this fact to quickly bring air to saturation through expansion cooling. In this lecture, we will see 2 more ways of bringing air to saturation. We'll also encounter 2 new kinds of temperature that help us understand atmospheric moisture. But, a few questions to consider along the way: What is black ice and when and where does it form, and can ice really be black? Speaking of ice, why are ice cubes you make in your freezer so often cloudy in the center? And why does it take longer to cook food at higher elevations?

So let's recap our 2 moisture properties that we learned about in the last lecture, vapor supply and vapor capacity. Vapor supply is the amount of water vapor in air, expressed as grams of water vapor per kilogram of dry air. Vapor capacity is also expressed in terms of grams per kilogram, but it tells us how much water vapor air could hold at a maximum. The ability of air to hold water vapor is a very strong function of temperature. Actually, we know now that it's the temperature of the water molecules that's actually relevant, but we're assuming that water vapor has the same temperature as nitrogen, oxygen, and the argon that make up most of the mass of the atmosphere.

So our concern in this lecture is bringing air to saturation, and we've seen one way already, lifting air in the dry adiabatic process. This relies on the fact that air is compressible and expandable, and the fact that pressure decreases with height. So as we force air to rise the external pressure pressing in on air decreases, and as the external pressure relaxes, expansion is permitted. And expanding air cools, and as the air temperature cools, the vapor capacity drops. As a result, we can take air with even just a little bit of water vapor, and we can lift it to the point where that air can no longer hold the water vapor it has. At that point, the relatively humidity is 100% and saturation has been achieved. We call this the dry adiabatic approach to saturation. There was no heat exchange, just expansion.

There are 2 other ways of saturating air. These introduce 2 new kinds of temperature, the dew point temperature and the wet bulb temperature. Actually, our familiar temperature is also known as a "dry bulb temperature" because it's measured with a thermometer bulb that is not wet. We'll see what makes the wet bulb wet pretty soon, but first, I want to talk about the dew point temperature, and I'd like to do that via a story.

It was a hot and humid summer day. I took a soda can out of the fridge and I set it on the kitchen table. After a few minutes, the can was covered in a glistening beading sweat. My son Ty, who was maybe 4 years old at the time, was positively delighted by this. And then he asked, "Where did this liquid water come from?" I told him that the soda can is riddled with thousands of tiny little holes, and as the can warmed up, increasing pressure inside the can forced the soda out through those holes, making it appear to sweat. My son knew that I like to pull his leg, and he said he didn't believe me, and I asked him, "How would you test my claim?" And he thought about it for a little while, and you can see the little gears moving in his head, and then he did something that I thought was really smart. He took his finger and he ran his finger along the outside of the can, and he tasted the fluid. And he said, "It's not soda."

I admitted that he was right, and then I said that the real story is the air is filled with tiny bits of water that are too small to be seen. But when they come in contact with the cold can, they catch a chill and this helps them grow large enough to become visible. My son looked at me for a while and then he said, "Nah." But actually, that's the true story, and what he had discovered was dew and the concept of the dew point. As we have seen, the regular, or dry bulb, temperature tells us the vapor capacity of air. Let's do an example. For air at sea level, if the temperature is 20°C, 68°F, the vapor capacity is 28 grams per kilogram. But if the temperature is cooler, 10°C, 50°F, the vapor capacity is only 15 grams per kilogram. The dew point temperature tells us the air's vapor supply. For air at sea level, if the vapor supply is 15 grams per kilogram, then it's the air's dew point temperature that is 10°C. Do you see? I'm using the same information we used previously. The temperature tells me the vapor capacity. The dew point tells me the vapor supply.

At saturation, vapor supply and capacity are the same, so the temperature and the dew point are also the same. So let's discuss the dew point approach to saturation, which is our second way of bringing air to saturation. The adiabatic approach, which we discussed previously, brought air to saturation without change of vapor content, by extracting heat. And we did that by expanding the air, by bringing it to a lower pressure. We can also bring air to saturation without changing its vapor content and without changing its pressure by extracting heat. This is a diabatic process. We're extracting heat and as a result, the air's temperature decreases. And when it decreases all the way to meet its dew point, we have reached saturation, the dew point approach to saturation.

Let's consider a subsaturated sample of air, and let's subject it to isobaric cooling. We're going to extract heat without pressure change and without altering its vapor supply. So suppose I have a temperature and I have a dew point. The temperature reflects the vapor capacity of the air, how much water vapor I can get in that air. The dew point temperature tells me vapor supply. Since the temperature is larger than the dew point, the supply is less than the capacity. But I'm going to cool this air now. I'm going to cause its temperature to decrease. How am I going to do this? I can do it in many ways. I can put the sample over ice. I could stick it in the refrigerator. Many, many ways of doing this, but the point is as the temperature decreases, the air's vapor capacity decreases as well. So its relative humidity, the ratio between supply and capacity, is increasing.

When I've managed to cool the air all the way down to its dew point, the sample is saturated. The sample is now saturated. The temperature and the dew point are the same. Supply equals capacity. Relative humidity, 100%. Any bit of extra cooling now makes it supersaturated, and condensation will appear as a result. Dew has formed. So the dew point approach to saturation—isobaric cooling without change of vapor supply.

Here's an example of air being cooled to its dew point. Let's start on a pleasant afternoon. I'm showing you temperature as a function of height, and how they might vary on this afternoon over Stick Man's height. Note the temperature is larger than the dew point everywhere, so the air is subsaturated. But let's say after sundown, the vapor supply of air doesn't change, but the vapor capacity decreases as the temperature of the air near the ground cools. Why? Because the ground's surface under Stick Man's feet has a low thermal inertia, and it's cooling quite well. That causes the air near the ground to cool, and the cooling is largest right near the ground.

If the air can cool all the way to its dew point, dew will form. Now, as we discussed in the last lecture, condensation requires a nucleus, a surface to actually condense on, but there's no shortage of those, and the ground surface is a great condensation nucleus. If the dew point were below freezing, then frost would form instead because the surface is also a pretty good ice nucleus as well. In this case, we would call it the "frost point," but it's basically the same idea. So let's talk about nighttime cooling.

The most likely place to find dew or frost is at the ground, since the ground typically cools better and faster than the air. The most likely time to find dew is around sunrise, since under most circumstances, that's the coolest time of the day. The most likely place along the ground to find dew is on objects that have low thermal inertia and are good radiators of energy—concrete, asphalt, metal, metal objects like railings and soda cans. Other good places are objects that have relatively large surface area for their size so that they can radiate away their heat overnight even more efficiently, in more directions, like blades of grass and your car windows.

You've probably seen signs that say, "Ice forms on bridge first." In that case, you're seeing a combination of the factors that we have discussed so far. Bridges are made of materials with small thermal inertia. They have more exposure to air, radiating energy away on all sides, especially below. Thus, other factors being equal, the bridge will cool to the dew or frost point before the adjacent roadway will. So this brings us to the topic of black ice. A particularly dangerous kind of ice can form on roadways, particularly bridges, when subfreezing air is chilled to the dew or frost point, but black ice is not black. It's actually a very thin layer of very transparent ice. It only looks black because it's taking on the color of the underlying road.

The 2 dangers of black ice are first, it is difficult to see because it blends in with the road because it is so transparent, and black ice can be especially slippery. Black ice is transparent in part because of its thinness. As water freezes, dissolved air is squeezed out, but oftentimes, some air is trapped and frozen, and trapped air bubbles is what tends to make ice cloudy. Your freezer's ice cubes are often clear at the edges, but have a cloudy center. Why is this? The reason is because those ice cubes tend to freeze from the outside in. The air in the outer edges of the water, as it freezes, can escape to the surrounding atmosphere because it's first to freeze, but the dissolved air in the water in the middle of the cube has nowhere to go, so it is trapped. Dissolved minerals can also make ice opaque, but for thin black ice, transparency is achievable and very dangerous.

Black ice is particularly slippery when newly formed. That means the temperature is close to freezing. Many of us, I certainly do, have painful empirical experience that ice is slippery. And amazingly enough, exactly what makes ice slippery is still a subject of debate, but one thing is clear. If a thin surface layer of liquid water develops on that ice, it is especially slick, and you can melt a thin layer of thin black ice by friction, by walking on it, by driving on it, and that's easiest to accomplish when the temperature is already close to the melting point. If air were able to cool to the dew point over a relatively deeper layer, not just at the surface, then we would have a fog instead. A fog is nothing more than a cloud with its base very near or at the ground.

I'd like to give you a forecasting rule, which is based on the information that we've discussed so far. Under the right conditions, air near the surface can chill to its dew point overnight fairly easily. Particularly if that surface has a low thermal inertia, if the winds are calm, the sky is clear, and the moisture content of the air aloft is not too large. Remember, water vapor is a greenhouse gas. Therefore, dew is not uncommon, particularly in deserts and indeed, nighttime dew is a major source of moisture for vegetation in places that otherwise don't get much rain and not much moisture. Once the surface air reaches its dew point, however, further cooling is a lot more difficult. Why is that?

The extra cooling would make the air supersaturated, and the excess vapor would then need to condense, and condensation is a warming process. When vapor becomes liquid, heat is released. We call this "latent heat," hidden heat. "Latent" comes from the Latin to lie hidden, the Greek word to escape notice. This heat warms up the air that holds the condensing vapor, and that's the air that was cooling. So this represents resistance to further cooling.

Here's the forecasting rule. If you don't expect your air mass at your location to change overnight, your minimum temperature will probably be no lower than your afternoon dew point. If the source and the nature of the air are not changing, the vapor supply is probably not changing very much either. Thus, the overnight dew point will probably not change very much, and we may not manage to cool air all the way down to the dew point, and many nights, dew does not form, even on the most favorable surfaces, but if we do manage to cool all the way down to the dew point, if we manage to cool all the way to that point, then further cooling will be more difficult because now there's resistance to cooling in the form of condensation warming.

If the air mass changes, if the sea or land breeze blows by, if a storm moves through, a front passes, etc., this rule doesn't work. But this rule helps illustrate the importance of water phase changes, a crucial source of heating and cooling. So let's talk about evaporation.

Condensation is a warming process. As we discussed, latent heat is released by the condensing vapor to the surrounding air. Evaporation, in contrast, is a cooling process, and of the 2, I think this one is probably more intuitive. When you step out of your shower, or even better, out of a swimming pool, you feel cold, especially if the wind kicks up. Why is that? One reason is because you're warmer than the air and the water on your skin, and water is a pretty good conductor. But actually, you didn't need to climb out of the swimming pool to feel that chill. The major reason is because the liquid water on your skin is evaporating into the subsaturated air. It takes energy to break the bonds that hold water molecules together into liquid water drops, energy in the form of heat, and the heat required to break those bonds comes from you.

This is, of course, why we sweat. Our body gives up some of its precious water in the hopes of gaining some evaporation cooling. So what does the wind in this situation do? The wind hastens the evaporation, and the cooling, by sweeping away those newly created water vapor molecules from the liquid water that evaporated from your skin. The water is evaporating since the air is subsaturated, and the vapor supply is less than the vapor capacity. That means there aren't enough vapor molecules to replace the molecules that have been escaping from the liquid water on your skin. In other words, the vapor pressure is too low, less than the saturation value.

As evaporation proceeds, the number of vapor molecules around your skin increases, and the vapor supply and the vapor pressure both get larger. In this way, the vapor deficit, the difference between the actual vapor pressure and its saturation value, the difference between the vapor supply and the vapor capacity gets smaller, so evaporation and the cooling it produces slows down. Does this seem familiar to you? It sounds a lot like wind chill. So how do you increase the evaporation rate? One way is to take all those newly created vapor molecules hanging around your skin and blow them away, so the vapor deficit is again large. You can push them away with the wind.

Now, let's talk about evaporation and the wet bulb temperature. We have seen that evaporation also moistens the air that it cools, and thus, it can bring air to saturation. Saturation by evaporation cooling, and moistening that results from that, is the wet bulb approach to saturation. Let's consider an example. Suppose I have a subsaturated air sample, temperature 30°C, 86°F, and the dew point is 10°C, or 50°F. The vapor capacity of this air, 28 grams per kilogram. The vapor supply, only 8 grams per kilogram. You're 20 grams per kilogram short of saturation. You want to saturate the air and this time, you choose to do it by supplying the missing vapor. Where is the vapor going to come from? Why don't you evaporate some liquid?

Can you fit 20 more grams of water vapor into that kilogram of air? The answer is no. Evaporation does increase the vapor supply, but evaporating that water takes energy from the air cooling it, and as it cools, the vapor capacity decreases, so the air is progressively less able to hold the water vapor you're trying to give it. Can I draw an example that you might see in your kitchen, or you won't see in your kitchen? It's as if you had a juice glass, and as you poured juice into that glass, the glass was actually shrinking. You'd get a lot less into that glass than you expected. The wet bulb temperature tells us what we should be expecting.

So let's start with subsaturated air. The vapor capacity is larger than the vapor supply. The temperature is higher than the dew point. And I'm going to evaporate some water into this air. As the water vapor content increases, the vapor supply increases and therefore the dew point goes up. But the temperature and the vapor capacity are decreasing. They're decreasing due to the cooling. The cooling is due to the evaporation of the liquid water that I'm using to raise the dew point. So as the dew point rises, the temperature decreases, and the temperature and the dew point are going to meet somewhere in the middle at a new temperature, and that's the temperature we call the "wet bulb."

For air brought to saturation by this process, the new temperature and the vapor supply are larger than their original values, but the temperature and the vapor capacity are also smaller. You didn't fit as much water vapor into the air as you were expecting, and that's the wet bulb approach to saturation. This also shows us the limitations of sweating. Here's another way of looking at the dew point and wet bulb approaches to saturation.

Let's suppose we have a temperature and a dew point and we have subsaturated air. If I subject the air to isobaric cooling, I can cool it all the way down to its dew point. Evaporation gets me less cooling because the temperature will only decrease to its wet bulb because the dew point is rising as I increase the vapor supply. So there's a limit to the cooling that I can get from sweating, particularly on hot, damp days that might be most oppressive to us. The best thing is to get some airflow going so you're not constantly trying to moisten the very same air. That's probably why you will fan yourself.

So let's do a very quick recap of our 3 temperatures. Any sample of air has 3 temperatures to describe its vapor properties: its regular or actual temperature, its dew point temperature, and the wet bulb temperature. All 3 temperatures tell us something important about humidity. The actual temperature tells us vapor capacity. The dew point tells us vapor supply, the actual water vapor content, and it shows how much I can cool the air without changing the vapor supply before saturation is reached. If I do this, the actual temperature also changes, becoming the same as its dew point, which is fixed in this particular process. The wet bulb tells me how much I can cool that air by evaporation. If I cool the air this way, its actual temperature is decreasing. It's changing and becomes the same as its wet bulb.

The dew point temperature of this air also changes during this process as well. At saturation, all 3 temperatures are equal and it doesn't matter how we got to saturation. At saturation, all 3 temperatures become the same, and this makes sense. If air is already holding all the vapor it can, it's already at its dew point because dew has already formed. If the air is already saturated, I can't evaporate any more liquid water into it, so it's already at its wet bulb. If the air is subsaturated, the wet bulb temperature falls in between the other 2 temperatures. The temperature is highest, the dew point is lowest, the wet bulb is in between.

We measure dew point temperature with a chart of the kind that I've been using, but how do we measure the wet bulb temperature and why is it called the "wet bulb temperature" anyway? So let's talk about that now. We measure wet bulb temperature with a device called the "sling psychrometer." This consists of 2 thermometers that are mounted side by side. One thermometer bulb is fitted with a sock, which we make moist by dipping in water, and this is what makes the wet bulb wet. So this is the wet bulb thermometer, and it will measure the wet bulb temperature. The other bulb is left dry in order to measure the actual temperature. So then we operate this thing by basically swinging it around. Why am I swinging this device? It's not changing the temperature that is measured by the dry bulb, but the device is being swung to force air through the sock, promoting evaporation of the air that is trapped between the sock and the wet bulb. And that is the sling psychrometer and how we measure the wet bulb temperature.

So now let's talk about boiling. You can make ice cubes in your freezer much more transparent, like the black ice on the roadway, if you boil the water first. Boiling purges dissolved air and minerals from the water that might make the water cloudy as it freezes. What is boiling? Boiling is actually very rapid evaporation, rapid evaporation from liquid to vapor state. We, of course, associate boiling with bubbles, and in fact, the word "boil" comes from the Latin word to bubble. What are these bubbles and why do they form?

Let's picture a pot of water on the stove. We have liquid water is overlain by air, and the air pressure is pushing down on the liquid water. Bubbles may form in the liquid water. What's the fate of a bubble of water vapor that might form in this pot? Atmospheric pressure is pushing down and also liquid water is pushing down on the water vapor bubble. If the pressure pushing down from both of those sources are larger than the vapor pressure pushing back, then the bubble will collapse, and it will not last very long. So how do you make a successful water vapor bubble? Success comes with higher temperature. The bubble is composed of water vapor at saturation, so the relevant vapor pressure pushing back in the bubble is saturation vapor pressure, and we know this increases exponentially with temperature. So the boiling point is the temperature at which saturation vapor pressure pushing out from the bubble matches the external pressure pushing back.

This is why boiling point decreases with elevation above sea level. Pressure decreases with height. As we ascend, there is less force pushing down, not only on us, but also on the bubbles. So the temperature at which saturation vapor pressure in the bubbles is able to stave off this external pressure pushing down is relatively lower. So therefore, this is why food takes longer to cook at higher elevation—because the boiling point to the water is lower, due to the lower atmospheric pressure. This illustrates, actually, a very important point. Phase changes occur at constant temperature, as well as constant pressure.

Suppose we heat a pot of water on the stove at a constant rate, and suppose we measure its temperature as a function of time, and we plot it on a graph. We would see the temperature increasing with time as we heat up the water, but it stops increasing when the boiling starts. During boiling, liquid water is being converted to vapor, but wait. I'm still adding heat to the water, so why isn't the temperature rising? The reason is because the energy being added to my system is not being used to increase the temperature of the water. Instead, it's going to breaking the bonds between the molecules in the liquid. Only after all of the liquid has been vaporized will the temperature of the water substance begin rising again. The phase change between liquid and vapor occurred at constant temperature.

Apply this to your food preparation. Let's say you're cooking noodles by boiling them in water. Heat from the boiling water is being absorbed by the food, principally by conduction. If the water boils at a lower temperature, the food cannot cook as quickly because the water won't be as hot, and you cannot make the water hotter by turning up the heat. All you can do is make the water boil faster.

I'd like to talk about the latent heat of fusion. The phase transition between solid and liquid water also involves latent heating and cooling. This is called the "latent heat of fusion," or melting, depending on the direction of the process. The energy required to break bonds between molecules and ice is not as large as that's needed for water. The latent heat of melting is 7.5 times smaller than the latent heat of vaporization, which is required for liquid water to convert to vapor.

So now let's summarize. In this lecture, we saw that there are 3 ways of bringing air to saturation, all 3 involving cooling. Air is subsaturated when its vapor supply is less than its vapor capacity, so saturating air is a matter of decreasing that capacity, increasing the supply, or doing both simultaneously. Vapor capacity is a very strong function of temperature. The very best way to decrease capacity is to lower that temperature. The adiabatic expansion approach to saturation operates by expansion cooling. The temperature drops as the pressure is reduced, without change of vapor content, until the vapor capacity and supply are the same.

Our second way, the dew point approach to saturation, involves extracting heat at constant pressure. We call that "isobaric," until the vapor capacity equals the vapor supply. The temperature at which the parcel would be saturated as a result of this isobaric cooling is called its "dew point." For the wet bulb approach to saturation, evaporation from liquid water provided the cooling, and this caused the vapor supply to increase. So the wet bulb temperature is reached when evaporation, cooling, and moistening have brought the air to an intermediate temperature between the original temperature and dew point.

We also saw why boiling point decreases with elevation. Boiling is very rapid evaporation. It was a matter of 2 pressures, atmospheric pressure pushing down, vapor pressure in bubbles pushing back. We also looked at what happens during phase transitions, including evaporation and condensation. Latent heat is exchanged when water substance changes phase. Latent means hidden, but the heat doesn't stay in hiding. When water vapor condenses to liquid, this heat is released to the surrounding air. When liquid needs to evaporate, this heat is taken back and stored. It becomes latent within the vapor molecule.

The water molecule is actually a very efficient carrier of heat energy. This latent heating can be carried a long, long way. In fact, one of the principal roles of clouds is to transport heat vertically, with latent heat release fueling potentially powerful updrafts. Whether clouds can accomplish this, however, depends on how stable the environment is. We'll start examining clouds and stability in the next lecture.

Lecture Nine
Clouds, Stability, and Buoyancy, Part 1

Scope:

In this lecture, we begin a detailed examination of clouds. We'll introduce the useful concept of the air parcel and learn 3 lapse rates, which quantify how temperatures decrease, or lapse, with height. Lapse rates will ultimately tell us if clouds are positively or negatively buoyant. During this lecture, think about why the bitterly cold air of the upper troposphere isn't exchanged with warm air at the ground.

Outline

I.   Clouds are dynamic and come in many shapes and sizes.

    A. Mounded clouds with substantial vertical development are called cumulus clouds.

    B. Lenticular clouds (lens-shaped) sometimes look like flying saucers and result from lifting moist air near mountains.

    C. Rotor clouds are the end result of violent downslope winds.

II. Clouds are classified with respect to the height of the cloud base—low, middle, or high.

    A. "Strato" is the first prefix used to indicate cloud base height. Stratus are low clouds, spread out and straight. "Alto" is the prefix used for middle-level clouds, and high clouds are cirrus, or "cirro."

    B. Two suffixes indicate the degree of vertical development. These prefixes and suffixes can be combined to describe clouds: cirrostratus, altocumulus, and so on.

    C. The recipe for clouds has 2 ingredients: moisture and phase changes, along with atmospheric stability and instability.

III. Let's consider a parcel of dry air.

    A. We assume that it has flexible sides, so the inside pressure adjusts as the outside pressure changes. We also assume that the parcel is closed and insulated; it experiences no mass or energy transfer with the environment.

    B. If a parcel is dry, closed, and insulated, there's no way to change its temperature except by changing the outside pressure.

IV. A lapse rate quantifies how temperature changes with height. Our expectation is that temperature decreases with height, resulting in a positive lapse rate.

    A. If we lift a subsaturated air parcel, it cools at a rate of 10°C per kilometer solely due to expansion. This is the dry adiabatic lapse rate (DALR).

    B. We saw that temperature tends to decrease significantly with height in the troposphere. This environmental lapse rate (ELR) is, on average, 6.5°C per kilometer in the troposphere. ELR varies in both vertical and horizontal space and in time.

V. Let's return to our dry air parcel.

    A. How temperature varies inside the parcel depends on expansion cooling, the constant DALR. How temperature varies outside the parcel is governed by the variable ELR.

    B. Consider an air parcel that is 15°warmer than the outside air. It's the same pressure as its surroundings but much less dense. It is positively buoyant and wants to rise. As it rises, it undergoes expansion cooling at the DALR, 30°F per mile.

    C. At 1 mile up, the parcel has cooled to 75°F in an environment of 70°F. By 2 miles up, the parcel is colder than its environment: inside temperature, 45°F; outside temperature 50°F.

    D. Up to this point, what's taking place is free convection (heat transport by fluid motion). Further ascent can only occur if the parcel is forced—forced convection.

    E. If the parcel were forced to rise much farther, it would become much colder and much denser than its surroundings. In reality, the parcel will likely remain in the environment where its inside and outside temperatures match.

VI. How can nature share all the hot air near the surface with the bitterly cold tropopause, just a few miles above?

    A. We've seen that we can't lift hot air very far; it doesn't stay hot. This illustrates an important concept called atmospheric stability—resistance to vertical displacement.

B. If we loft a parcel of air, its density increases at the same pressure, and it becomes negatively buoyant. If we push the parcel down, its density decreases at the same pressure, and it becomes positively buoyant. This situation is called absolutely stable.

C. The airplane example helps illustrate why the troposphere doesn't turn over. If we drag the tropopause air down to the surface, it gets much hotter, much less dense at the same pressure, than the air that's already there.

VII. Nature's way of lofting surface air to the tropopause is thunderstorms.

A. Rising air expands and cools because pressure decreases with height (dry adiabatic process). The relative humidity of rising moist air also increases. That's how we created a cloud from lifting because vapor capacity decreases quickly as temperature goes up.

B. Further lifting of saturated air means further expansion cooling, but now there's a vapor excess (supersaturation). Some of that excess vapor must condense, and condensation is a warming process. What receives the warming? The air in the supersaturated parcel.

C. The net result is that the cooling rate for the parcel on ascent is cut in half, to 5°C per kilometer. This is the moist adiabatic lapse rate (MALR). Unlike the DALR, the MALR is not a constant.

D. Let's look at a saturated parcel of air. It starts at the same initial temperature as its environment, 30°C, and holds 28 g/kg of water vapor. If we push that parcel up 1 kilometer, it cools at only 5°C.

E. The parcel is now 1°C warmer than its new environment and positively buoyant. At 2 kilometers, the saturated parcel is now 2°C warmer and rising faster.

F. The ELR, usually 6°C per kilometer, is actually larger than the MALR, 5°C per kilometer, so the rising saturated parcel can become warmer than its environment. And the farther it rises, the more its temperature increases in relation to its environment.

G. The reason the atmosphere isn't always in a state of chaotic mixing is that we don't encounter saturated air near the surface very often. Air typically starts off subsaturated and cooling faster than the ELR before finally reaching saturation.

Suggested Reading:

There is no suggested reading for this lecture.

Questions to Consider:

1. You are driving in your car and suddenly your windows start fogging up from the inside. Is it better to turn on the defrost heater or the air conditioner? What does each device do to the car interior's air?

2. On very humid days, we often say the air "feels" heavy. Is moist air denser than dry air, at the same temperature? Hint: Adding water vapor to the air increases the value of the "gas constant" r in the ideal gas law: $p = \rho r t$, where p = pressure, ρ = density, t = temperature, and r is the gas constant.

Lecture Nine—Transcript
Clouds, Stability, and Buoyancy, Part 1

Welcome back to Meteorology, our survey of the wonders of the weather. You'll recall that we have 3 temperatures, our regular dry bulb temperature, our dew point temperature, and the wet bulb. Each tells us something about moisture, such as vapor capacity, vapor supply, and how much we can chill air through evaporation. If we want to bring unsaturated air to saturation, cooling is involved. We can saturate air through expansion cooling, by evaporating liquid into it, or by extracting heat with an air conditioner or something like that. Cooling air reduces its vapor capacity.

In this lecture, we commence a detailed examination of clouds, how they come about, why are they thick or thin, long-lived or short. We'll introduce the very useful concept of the air parcel, a blob of air that we'll follow around and monitor its temperature and humidity. We'll have 3 lapse rates to keep track of. Lapse rates quantify how temperatures decrease, or lapse, with height. Lapse rates will ultimately tell us if clouds are positively or negatively buoyant. Positively buoyant clouds grow into the kind of deep clouds associated with severe weather.

Some things to think about during this lecture: We saw that we have warm air at the ground and bitterly cold air at the tropopause, just a few miles up, and yes, we saw that density decreases with height, and that's why the troposphere is not turning over. But why don't we give some of this warm air to the upper troposphere, and why don't we just grab some of that cold air down and bring it to the surface? After all, they can do it in the movies.

We've been working hard on our moisture concepts, so let's start off by looking at some cloud pictures. I became a meteorologist because of clouds. Probably a lot of us did. I still study clouds, study the way they form and die, and how they move. My field is called "cloud dynamics" because clouds are dynamic. They're always changing. Clouds can be very beautiful in their own right, and for the things that they can do with light. Many times, clouds are very benign and even do appear like the proverbial ice cream castles in the sky. The cloud particles reflect all of the colors of visible light to our eyes, which is why they usually appear white. A common sight are these rays from heaven, which are called "crepuscular rays." That's a weird-sounding word, but it comes from the Latin word for dark, "crepuscular," implying dusk or twilight. But these rays can occur any time that light is channeled through gaps in objects, such as clouds.

Here are crepuscular rays heading skyward, making an angelic halo around a cloud. I took this picture. I was just walking down the street one day and I happened to look up. It's one of my luckiest shots. Well, actually, maybe I wasn't so lucky. I look up a lot. That's sort of an occupational hazard. This is a picture of Lake Tahoe at sunrise. The lake appears to be on fire from reflecting the rising Sun's light, coming down from the clouds above. Many photographers realize that what really makes a picture is how clouds manipulate the light.

Clouds come in many shapes and sizes. Mounded clouds with substantial vertical development are called "cumulus clouds," from the Latin term to pile up, to accumulate. These lens-shaped, or lenticular clouds, often result from lifting moist air near mountains. Lenticular clouds can look like flying saucers, and sometimes they're even mistaken for them. But not all clouds associated with mountains are quite so benign. These rotor clouds downwind of California's Sierra Nevada Mountains are the end result of violent downslope winds. We'll look at why these winds form later.

Clouds can take curious shapes. Here's a kind of water wave that cannot be surfed. These waves result when the vertical wind shear is strong, and we'll talk about shear and its role in cloud development later. Speaking of oceans in the sky, this strikes me as a school of dolphins, swimming in formation. I wonder what that says about me. Band-like formations of clouds are very common, one reason being there are several different mechanisms for organizing clouds in this fashion. We'll see more examples of this later in the course.

Clouds are dynamic. Nobel Prize-winning physicist Richard P. Feynman once said this about clouds, "As one watches them, they don't seem to change, but if you look back a minute later, it is all very different." We can capture this change through time-lapse photography. Here is perhaps 30 minutes in the history of a storm, compressed into a few seconds. Note the storm is composed of several short-lived elements. We call these "cells." They seem to inflate like balloons, reaching for the sky, only to collapse. In long-lived storms, they get replaced by new balloon-like cells. This storm is a long-lived collection of short-lived cells.

Here are clouds perched on a mountain range, or are they really perched? If you just glanced at the mountains for an instant, you might think that the air was still, and the cloud was just sitting like a hat on someone's head, but in fact, the air could be moving, and moving quite quickly. The cloud could be bubbling and writhing, as in the case here. I'm picturing a giant white hand, its fingers desperately crawling over from the other side of the mountain, blindly grasping for something on our side. What is it grasping for? Well, I'm not sure.

Here's a thunderstorm anvil cloud, darkened by the presence of large precipitation particles. To the right, a precipitation shaft can be seen where the rain is particularly intense. Probably hail is falling there as well. To the left, the ominously lowered wall cloud, where tornadoes often appear. Beneath thunderstorm anvils, this eerie bulbous formation called "mammatus" sometimes appears. It isn't a thunderstorm without thunder and lightning, which, in a sense, is clouds producing their own frightening and beautiful light.

Feynman also had this to say, "Anyone who has been in a thunderstorm has enjoyed it, or has been frightened by it, or at least has had some emotion. And in those places in nature where we get an emotion, we find there is generally a corresponding complexity and mystery about it." We're going to learn about that mysterious complexity, but we will need more tools in our toolbox first. Let's start with some terms, the names we use to describe and classify clouds.

We classify clouds with respect to the height of cloud base—low, middle, or high. The 3 prefixes that we use to indicate cloud base height are first, "strato," as in stratus. That means spread out and straight. We've seen strato already in the word "stratosphere." Stratus are low clouds. Second, "alto," which actually is the Latin word for high, but we use it to mean middle. The reason is because our high clouds are "cirrus" or "cirro," from Latin to curl. These are our high, wispy, and feathery ice clouds. We also have 2 suffixes, indicating the degree of vertical development, sheet-like "stratus" or lumpy "cumulus." We can combine our prefixes and suffixes in many different ways, so we can have "cirrostratus," "altostratus," "cirrocumulus," and "altocumulus." We can even have "stratocumulus," which seems like a contradiction, but actually, those are among the most common clouds on Earth, sheet-like clouds, punctuated with lumps.

We can have subtypes as well. An example of altocumulus clouds is our lenticular clouds produced by mountains, altocumulus lenticularis. The tube-like roll clouds I showed you are stratocumulus undulatis, undulating lumpy sheets. Cumulus humilis are our fair-weather cumulus clouds. Humilis means humble. When clouds are heavy with rain, we insert the word "nimbo" in the name. Nimbostratus clouds darken the sky in sheets, and cumulonimbus—that's our thunderstorm cloud. Now that we have names for our clouds, let's return to our cloud recipe, which has 2 ingredients. Moisture and phase changes— we covered that in the last lecture, and atmospheric stability and instability—we start on that now with the introduction of the air parcel concept. I already mentioned to you that if we take a blob of air and force it somewhere where the pressure is lower, the air expands. As a result of this expansion, the air cools and if it carries water vapor, its relative humidity rises. We can saturate moist air this way. This was the dry adiabatic approach to saturation. In this thought experiment, we were making use of the air parcel concept. We followed a sample of air around, monitoring how its properties, like temperature and relative humidity, changed, but we need to make some assumptions, typically, when we follow air parcels around.

So let's consider a parcel of dry air for now. We'll add water vapor later. The typical assumptions and simplifications we have to make are first, we assume our air parcel has flexible sides, so that the inside pressure adjusts as the outside pressure changes. So as an example, I make a parcel near the ground, where the pressure is 1000 millibars, and the inside pressure in my parcel is also the same. Then I lift my parcel to somewhere where the pressure is 800 millibars. My inside pressure drops to match the new outside pressure. How? By permitting the air to expand.

Second, we usually assume that the parcel is closed and insulated. "Closed" means no mass transfer with the environment, by which I mean its surroundings, no mixing. "Insulated" means no energy transfer with the environment, no absorption or emission of radiation, no conduction. This is important. If a parcel is dry, closed, and insulated, there's no way to change its temperature, except through expansion and compression, by changing the outside pressure. That is the dry adiabatic process. So I think we're ready for a very counterintuitive example.

It's a hot day. Let's go outside and make a parcel of that outdoor air, just dry air for now. It's 90°F, 1000 millibar pressure. The parcel is closed and insulated. Let's bring it indoors, where the temperature is 68°F, but the pressure is still 1000 millibars. My parcel is still 90°F. I have not changed its volume. So I go to the freezer, and I put my air parcel in there, where the temperature inside the freezer is 0°F and the pressure is 1000 millibars. My air parcel is still 90°F. How is that possible? The reason is the parcel is dry, closed, and isolated. We didn't change the exterior pressure, so the interior volume of the parcel did not change. No volume change, no temperature change. Now, I know what you're thinking. This seems ridiculous. Eventually, the parcel will cool off, right down to the temperature of the freezer, right? Yes, it will, but how long will it take?

If we check back on our parcel after a few hours, we'll find that the temperature of that air is definitely lower than 90°F. But what if we go back and check only after a few minutes? Actually, we'll be moving these air parcels around pretty quickly, so this is a better assumption than it might appear to be.

Let's switch gears to lapse rates. A lapse rate quantifies how temperature changes with height. Our expectation is that temperature decreases with height. It lapses, so that makes for a positive lapse rate. We have already seen 2 lapse rates. The first one is constant and fixed. If we lift a subsaturated air parcel, it cools 10°C per kilometer, or almost 30°F per mile, solely due to expansion. This is the dry adiabatic lapse rate, and I'm going to call it by its initials sometimes, "DALR." It describes how temperature changes with height, due to the dry adiabatic process.

We also saw that temperature tends to decrease with height in the lower part of the atmosphere called the "troposphere." That was because air doesn't absorb much incoming shortwave radiation, and it's also a pretty lousy conductor. Remember that temperature decrease was huge, 140°F temperature drop over the 7.5 miles between the surface and the tropopause. In metric, that's 78°C temperature drop over 12 kilometers. But in either system, that's our environmental lapse rate or "ELR." It checks in at 6.5°C per kilometer, 19°F per mile, in the troposphere.

In the stratosphere, the ELR is negative because we have an inversion. Temperature increases with height, so we're already seeing that the ELR varies in the vertical. It varies in horizontal space and in time as well. So suppose one day, the temperature profile looks like this instead. In the troposphere, I see 2 places where the lapse rate is a little steeper, larger than the standard 6.5°C per kilometer rate, and there's something that we call a "capping inversion" a few kilometers above the ground. This capping inversion could keep clouds from forming, and it also could mean that when clouds do form, they're explosive. I also see the tropopause as a little higher, the stratosphere a little colder in this particular example. The point is the ELR is as variable as the weather itself, but 6.5°C per kilometer, 19°F per mile, is a very reasonable average ELR for the troposphere.

Well, so what? We need to appreciate that the DALR and the ELR are 2 very different lapse rates. Ultimately, that difference will reveal 2 things to us—why we can get thunderstorms and why we don't get thunderstorms everywhere or every day. Let's start on understanding that. Let's lift a parcel. So let's make a closed and isolated parcel of dry air. We're still ignoring moisture for now, and let's lift it, the inside and outside pressure presumed the same. The parcel expands as the surrounding pressure around it eases. How temperature varies inside the parcel depends on expansion cooling, the constant DALR. How temperature varies outside the parcel is governed by the ELR, which can vary from place to place and time to time. Whether the parcel continues rising or not depends on the difference between these 2 lapse rates.

So when we lift a parcel, the extremely important question is, will it keep rising? Remember, less dense air rises, more dense air sinks. Also recall that temperature, pressure, and density are tightly related through the simple but powerful ideal gas law. At the same pressure, warmer air is less dense than colder air. We are presuming the parcel is the same pressure as its surroundings. This means that if the parcel is colder than its environment, it is also denser, and it wants to sink. So let's try our new concept out on the beach.

It's a warm summer day. You're lying on the beach, frying like a piece of bacon, looking up at that beautiful sky, and wondering why it's blue. I'll tell you later. And it dawns on you that as hot as you are, subfreezing temperatures are just a few miles above your head. Say on this day, the surface temperature is 90°F. Say the ELR on this day is 20°F per mile, pretty close to its long-term average. That means that 1 mile above you, the temperate is 70°F. By 2 miles up,

50°F. By 3 miles up, you're already below freezing. You leave the beach and you go to your dark-colored car, which has also been baking in the Sun. The temperature inside that car, 105°F. This air escapes when you open the door. You just made an air parcel.

This parcel is the same pressure as its surroundings, but it's 15°F warmer, therefore much less dense than its environment. It is positively buoyant and it wants to rise. As it rises, it expansion cools at the dry adiabatic lapse rate, 30°F per mile. Meanwhile, the surrounding environment is also getting cooler, but at the slower ELR. So 1 mile up, the parcel has cooled to 75°F. Its new surrounding environment is 70°F. Our parcel started with a 15°F temperature advantage over its surroundings, and it's already squandered almost all of it. It is only 5°F warmer, but it's still warmer and less dense than its surroundings, so it keeps on rising.

By 2 miles up, the parcel is now colder than its environment, inside temperature 45°F, outside temperature 50°F. Up to now, we have had what we call "free convection." Convection is heat transport by fluid motion, heat transport by parcel ascent. The convection was called "free" because it was driven by its own positive buoyancy. Now, further ascent can only occur if the parcel is forced, forced convection. If the parcel were forced to rise much farther, it would become much, much colder, much denser than its surroundings. By 4 miles up, the parcel would be 25°F colder. In reality, the parcel isn't going to get anywhere near there. It would likely have remained near the level in the environment where its temperature matched the outside temperature. We call that the parcel's "equilibrium level."

There's an important lesson here. Let me illustrate it with a story. The story involves a low-down, mean, dirty trick, so if you do this, don't blame it on me. We've spent a lot of time at the beach in this course, and you know, kids love playing on the beach. They love digging holes in the sand and then they love filling up those holes with water. Go down to the beach and give a kid a bucket, but give him a bucket with holes in the bottom. So the kid goes to the ocean or the lake and he fills up his bucket with water, and then he brings it back to the hole that he has painstakingly dug in the sand, and he realizes that his bucket is empty. So he tries it again, and he comes back. The bucket is still empty. All the water rushed out the bottom of the bucket. All of that nice water there, his dry hole here, and no way of getting the water from there to here.

Nature is also frustrated, just like that kid. All that nice hot air near the surface—how to share it with the bitterly cold tropopause, just a few miles above? But we've seen that you can't lift hot air very far. Hot air doesn't stay hot. We have just seen an important illustration of a concept called "atmospheric stability." Atmospheric stability is resistance to vertical displacement. Let's look at this in a slightly different way.

Here's the standard atmosphere again, with an average ELR of $6.5°C$ per kilometer. Let's make a dry air parcel out of environmental air at some level in the troposphere. By luck, 1 kilometer above or below the parcel, I would find air in the environment $6.5°C$ cooler or $6.5°C$ warmer than this new parcel. But if I displace this parcel vertically, its temperature changes at the DALR of $10°C$ per kilometer, so if I push up this air parcel 1 kilometer, it cools $10°C$. It is now $3.5°C$ colder than its environment. It's $6°F$ colder than its surroundings. It's more dense at the same pressure and negatively buoyant. It wants to sink back down where it came from.

If instead I displace it downward, it warms $10°C$ instead, due to compression. It's now $3.5°C$, $6°F$, warmer than its surroundings, less dense at the same pressure and positively buoyant. The parcel wants to rise back up where it came from. We push air up, it wants to go down. We push air down, it wants to go up. This atmosphere is stable to vertical displacements of dry air. This situation is called "absolutely stable." It also helps illustrate another powerful reason why the troposphere doesn't just turn over. To see why, let's get on an airplane.

You've probably noticed that aircraft aren't particularly well insulated. They're cold and noisy. They're not pressurized all the way to sea-level pressure either. Let's see why. Picture an airplane flying at the tropopause, 12 kilometers up, where the outside pressure is 200 millibars, outside temperature $-60°C$, $-76°F$. Let's say the cabin interior is pressurized to 1000 millibars, but it isn't perfectly sealed. Therefore frigid outside air has to be brought in to maintain cabin pressure. It's going to be very cold in that plane, right? No. Actually, the air conditioners on that plane are going to have to run full tilt, and it has nothing to do with the number of bodies on board that aircraft. The reason is because we need to compress air by a factor of 5, just to get it in the airplane. Outside pressure, 200 millibars. Inside pressure, 1000 millibars. And as we squeeze the air into the airplane, the volume changes and the temperature changes. It's going to get very, very hot.

This is equivalent to forcing the air to descend those 12 kilometers from the tropopause to sea level dry adiabatically. That causes the temperature to increase by 120°C, or 216°F, during that descent. The air parcel started very cold at −60°C, −76°F, but the new parcel temperature inside the plane, once it gets inside, is plus 60°C, plus 140°F, and all we did was change the volume by changing the pressure. That's a big reason why planes aren't pressurized all the way to sea level pressure, and the reason why they're not too well insulated either. This example also helps illustrate why the troposphere doesn't turn over. The troposphere represents warm air beneath cold air. We previously saw that warm air rises and cold air sinks is not always true. But if we take the tropopause air and we drag it down to the surface, we see it is much hotter, much less dense at the same pressure than the air that's already there, and this works in reverse as well.

I mentioned in Lecture One a scene in *The Day After Tomorrow*, the movie, where very cold air was brought down from the upper troposphere or stratosphere down to New York City. It reaches the surface, freezing, and it just flash-freezes everything. We've just seen that that can't happen. If we take air from the tropopause, or the stratosphere, and we bring it down to the ground, it is very, very hot.

We have gone as long as we could talking about clouds and stability, without considering the powerful role of moisture. Meteorology is not a dry subject. Nature does have a way of lofting surface air to the tropopause after all—in thunderstorms because water vapor is rocket fuel.

We've already seen that we can lift air to saturation. Remember, as water vapor condenses, heat is released to its surroundings. That heat is the fuel that we are talking about, so let's start contrasting subsaturated and saturated ascent. Rising air expands and cools because pressure decreases with height, our dry adiabatic process. Remember, "dry adiabatic" really means subsaturated. The air parcel can contain moisture. It just must have a relative humidity under 100%, and we've also seen the relative humidity of rising moist air increases as well. That's how we created a cloud from lifting because vapor capacity decreases quickly as temperature goes down.

Air is quickly becoming less able to carry the water vapor that it's carrying with it. Further lifting of saturated air means further expansion cooling, but now there's a vapor excess. There's more water vapor than the air can hold. This is supersaturation. Some of that excess vapor has got to condense, and condensation is a warming process. What's receiving the warming? The air and our supersaturated parcel. The net result is that the parcel will still cool on ascent, but not as quickly. Instead of 10°C per kilometer, or 30°F per mile, that cooling rate is going to be cut in half to 5°C per kilometer, to 15°F per mile. We call this the "moist adiabatic lapse rate" or "MALR." Unlike the DALR, the moist adiabatic lapse rate, MALR, is not a constant, but we don't have to worry about that right now.

So let's go back to the beach. My example now is using Celsius and kilometers instead. Here's my dry air parcel, which cooled at the DALR. It started off hot and positively buoyant, but it didn't stay that way. Now, let's make a saturated parcel. The same initial temperature as its surrounding environment, 30°C, it doesn't get the head start my dry parcel did, but it's carrying a lot of water vapor. Air of that temperature can hold 28 grams per kilogram. If we push that parcel up 1 kilometer, it cools only 5°C at the moist adiabatic lapse rate.

Compare its temperature to its new environment. It's 1°C warmer. It wants to rise on its own. At 2 kilometers up, our dry and saturated parcels have the same temperature, but while the dry parcel is running out of gas, the saturated parcel is just getting started. It's now 2°C warmer than its environment and it's rising faster. Let's jump ahead to 4 kilometers up. At 4 kilometers, the dry parcel is 6°C colder than the environment. If we pushed it there and let it go, it would drop like a rock, but the saturated parcel is 4°C, or 7°F, warmer. That may not seem much to you, but that air is probably rising at 40 mph.

So let me ask a question. What we're noticing here is the environmental lapse rate at usually 6°C per kilometer is actually larger than the moist adiabatic lapse rate at 5°C per kilometer, so that rising saturated parcel can actually become warmer than its environment. And the farther you raise that air parcel, the warmer over its environment it'll be. Then you might ask, why isn't the atmosphere always in a state of constant chaotic mixing? How often do you encounter saturated air near the surface? The answer is at the very tip of your nose—not that often. Air typically starts off subsaturated and cooling faster than the environmental lapse rate, before finally reaching saturation. In this way, air usually starts negatively buoyant, and that's an obstacle to overcome. Can it be overcome? Will it be overcome? Well, that makes this story much more interesting.

So let's summarize what we covered in this lecture. Through photographs, we saw that clouds can be beautiful or threatening. They can manipulate light and even make their own. We familiarized ourselves with some of the descriptive terms we use for these clouds. We made use of the air parcel concept, in which we identify a mass of air and follow it around as its temperature, humidity, and elevation change. How a parcel's temperature changes on ascent depends on whether or not it's saturated.

If the parcel is subsaturated, it cools at the rapid DALR, but we saw that we cannot lift hot, dry air very far at all. We also saw that we can't make cold air descend very far. This situation is called "stable," but when the air is saturated, condensation warming partially offsets expansion cooling, resulting in a substantially slower cooling rate, 5°C per kilometer, moist adiabatic lapse rate. The key difference between these 2 rates—the average environmental lapse rate 6.5°C per kilometer falls in between. We can lift hot saturated air a whole lot farther. Water vapor makes the atmosphere a much more interesting and complex place.

Let's look ahead. We will continue exploring the mysteries of clouds in the next lecture. We'll see why mountains cause clouds, why deserts form in the lea of mountains. We'll see the source of thunderstorm energy and why thunderstorms aren't more frequent than they already are.

Lecture Ten
Clouds, Stability, and Buoyancy, Part 2

Scope:

In this lecture, we continue our discussion of clouds, stability, and buoyancy. We'll see why it matters if condensed vapor remains in the parcel as cloud water or falls out as rain. We'll see why deserts are often found in the lee of mountains; why thunderstorms form; and why they don't form as often as they could.

Outline

I. We begin by returning to our flexible, closed air parcel and recalling the 3 lapse rates we've learned about: the ELR, DALR, and MALR.

    A. We've learned that we can't lift hot, dry air very far, but a saturated parcel of air will rise quickly. The state of the rising saturated parcel is called conditional instability, that is, conditioned on the presence of moisture and the situation of saturation.

    B. Rising saturated air parcels can be warmer than their environment, but we don't encounter warm saturated parcels at the surface very often. Instead, air typically starts at subsaturation, cooling faster than the ELR before finally becoming saturated.

    C. If we lift a subsaturated air parcel, its temperature decreases at the DALR. The dew point of the parcel also decreases at about 2°C per kilometer. That's the dew point lapse rate.

    D. The dew point decreases because the volume of the parcel increases as it ascends; the water vapor in the parcel spreads out.

II. A film clip of clouds in motion shows this dry adiabatic approach to saturation, followed by a moist adiabatic ascent above cloud base.

    A. At cloud base, temperature and dew point of a subsaturated air parcel are both the same and both lower than when they started.

    B. We now have a cloud, extending from the LCL up the mountaintop.

C. If the parcel is positively buoyant, it continues to rise and creates a deep convective cloud. If it's negatively buoyant (colder than its surroundings), it will sink back down the leeside of the mountain.

III. Now let's look at a thunderstorm sounding, a case in which it's possible for a surface parcel to become positively buoyant on ascent, despite starting off subsaturated and struggling.

    A. A subsaturated parcel lifted to LCL is negatively buoyant, but if a front pushes the air aloft, the now-saturated parcel will follow the moist adiabat and become warmer than its surroundings. It reaches its level of free convection (LFC) and is positively buoyant.

    B. Ultimately, the rising parcel will exhaust its supply of water vapor and cool down to the same temperature as its surroundings. This is its equilibrium level (EQL), also called cloud top.

    C. The main part of the cloud is warmer than its surroundings and positively buoyant. The lower section, between the LCL and LFC, is colder, as is the top part of the cloud, slightly above the EQL.

IV. Let's consider the region of positive buoyancy.

    A. The area from the LFC to the EQL represents potential energy (called convective available potential energy [CAPE]) that can be used by the parcel to rise.

    B. Before the CAPE can be tapped, negative buoyancy must be overcome. That's the negative area on our sounding, called convective inhibition (CIN).

    C. Our sounding has an LFC for surface air, and that air can become positively buoyant if lifted high enough, but it doesn't always have the means to reach LFC.

V. Let's return to the sea breeze, which often initiates thunderstorms.

    A. On the beach, the land surface heats up during the day, and the thickness of the layer between 2 isobars increases.

    B. Air rises over the land and spreads toward the sea, creating a pressure difference between land air and sea air. The cool sea air starts moving inland.

C. When the cooler air from the sea collides with the warmer air over the land, a front is formed.

D. The sea air burrows beneath the less dense air over land, forcing the land air to rise over the sea breeze front. If the land air is sufficiently moist and if the frontal lifting is strong and deep enough, the land air can be lifted to its LCL.

E. If the land air reaches its LFC, the sea breeze lifting can provoke deep convection, leading to strong winds, heavy rain, and possibly, thunderstorms.

VI. Roll clouds result from strong uneven heating of the land.

A. As the land heats up, the relative humidity decreases because, typically, the vapor supply does not change. That would make it seem harder to create a cloud by lifting.

B. We can bring a parcel of subsaturated surface air to saturation at 600 meters, but this parcel will always be negatively buoyant.

C. As the land surface heats up during the day, however, the ELR increases near the surface. At this point, the temperature and the vapor capacity have increased, but the dew point and the vapor supply have not; thus, the air now has a lower relative humidity.

D. To make a cloud from this surface air, we have to lift it even farther, but there's much less resistance to lifting. Even though it's easier to lift, our air parcel doesn't remain positively buoyant very long and will probably not get much above its LCL. The result is a string of tiny clouds.

Suggested Reading:

There is no suggested reading for this lecture.

Questions to Consider:

1. How might thunderstorms be different, in intensity and/or depth, if all of the atmosphere's ozone completely and permanently disappeared?

2. You heat a vessel full of water. After it starts boiling, you remove the vessel from the heat and set it aside. After the boiling motion ceases, you dip a metal spoon into the upper layer of the water. The water begins boiling again. Why?

Lecture Ten—Transcript
Clouds, Stability, and Buoyancy, Part 2

Welcome back to Meteorology. In this lecture, we continue our discussion of clouds, stability, and buoyancy. We will discovery why it matters if condensed vapor remains in the parcel as cloud water or falls out as rain. We'll see why deserts are often found in the lee of mountains, why deep powerful thunderstorms can form, and why they don't form as often as they could. This review will be a little longer than most because we're in the middle of a complex but fascinating topic. But we're into the home stretch now, so let's go.

We've been making use of the air parcel, a blob of air of unspecified size, but with flexible sides. "Flexibility" means if I increase or decrease the pressure outside the parcel, the inside pressure adjusts to match. The easiest way to change the outside pressure is to move it vertically. We usually also assume the parcel is closed and insulated from its surroundings. That means no heat transfer and no mixing. This idealization is more realistic than it might seem right now. Remember, a lapse rate tells us how temperature decreases or lapses with height.

We've had 3 different lapse rates so far; the ELR, DALR, and MALR. The ELR, or environmental lapse rate, varies in space and time, but it averages 6.5°C per kilometer in the troposphere. The DALR, dry adiabatic lapse rate, is fixed at 10°C per kilometer. The MALR, or moist adiabatic lapse rate, actually varies as well because it depends on how much water vapor is actually condensing, but we've been taking it to be 5°C per kilometer, and that's a pretty reasonable average.

We've been spending a lot of time on the beach in this course, so let's head back there again, armed with our 3 lapse rates. A few things to think about on our journey: Looking up from the beach, you've probably seen clouds overhead, shallow but seemingly organized, looking like pearls on a string, especially in the early to middle afternoon. Or you may have seen deep clouds form over your head and then pass rapidly inland. We'll see these clouds during this lecture.

So it was a warm day. The surface temperature was 30°C and the ELR was 6°C per kilometer. We made a parcel of dry air that started off scorchingly hot, but it didn't stay hot very long because it cooled at the very rapid DALR of 10°C per kilometer of ascent. We saw from this

©2010 The Teaching Company.

that we can't lift hot dry air very far. It just won't stay hot. But our saturated parcel took off like a rocket. We started with a surface temperature of 30°C, the same as its surroundings, but loaded with water vapor, all the way up to saturation. And as the saturated parcel ascended, it still expanded and cooled, but its vapor capacity also decreased, and that caused the air parcel to become supersaturated. The excess vapor had to condense, and that heat was released to the air in the air parcel, and that heat reduced the lapse rate in half, from the DALR of 10°C per kilometer to the moist adiabatic lapse rate of 5°C per kilometer, and that made all the difference.

Thus we saw we had a situation in which a rising dry air parcel never became warmer than its surroundings, or its ascent in the atmosphere was very limited. But a rising saturated parcel could and did become warmer than its surroundings. This is called "conditional instability." Conditional instability, conditioned on the presence of moisture and the situation of saturation. We had conditional instability because the moist adiabatic lapse rate is less than the typical environmental lapse rate in the troposphere. The MALR 5°C per kilometer, the ELR averaging 6.5°C per kilometer. But if rising saturated parcels can be warmer than their environment, then why isn't the atmosphere always turning over? And the reason is we don't encounter warm saturated parcels at the surface very often, not even on the beach.

Instead, air typically starts subsaturated, cooling faster than the environmental lapse rate, before finally becoming saturated. This air usually starts off negatively buoyant, and that's an obstacle to overcome. And on many days, it won't be overcome. So let's take a closer look. Let's raise a subsaturated air parcel. We know it has a temperature and a dew point, reflecting its vapor capacity and vapor supply. If I lift the parcel, its temperature decreases at the DALR. I also need to reveal now that the dew point of this parcel also drops at about 2°C per kilometer. That's the dew point lapse rate. Actually, that's our fourth lapse rate, and if I had told you at the beginning we would have 4 lapse rates to contend with, you might have given up on me, but we've made it to here, so let's keep moving upward.

So here's our subsaturated air parcel. Temperature is higher than the dew point and we're going to lift it. The parcel temperature cools at the DALR, since it's subsaturated, and it gets cold fast. I'll call this line the "dry adiabat," because it's the path of the dry adiabatic process. Its dew point is also decreasing, but at a much slower rate of

2°C per kilometer. The dew point is decreasing because our parcel is getting larger. Its volume is increasing, so the water vapor in the parcel is becoming more spread out. That makes collisions more rare, so we actually have to cool the air farther to compensate for all that extra room that the water vapor has to roam in.

But the crucial point is these 2 slopes are very different, and they're going to cross. And when they cross, the parcel is saturated. The relative humidity is 100%. The temperature equals the dew point. This is the lifting condensation level, or "LCL." That's the level at which condensation can be achieved by lifting. So we've accomplished the dry adiabatic approach to saturation. LCL has another name as well, "cloud base." We've reached cloud base, so what happens next? We're finally saturated. Now we cool at the moist adiabatic lapse rate, roughly 5°C per kilometer. I will call this path the "moist adiabat" because it's the path of the moist adiabatic process. Note I drew the moist adiabat as a curve, rather than a line. This is admitting that the moist adiabat lapse rate is not a constant.

I previously showed you this movie of clouds in motion over mountains. This is an example of the dry adiabatic approach to saturation, followed by a little moist adiabatic ascent above cloud base. Let's look at this process in detail. Consider a mountain. The wind is blowing from left to right. Let's make a parcel of surface air and see what happens as it approaches the mountain. The air is forced to ascend. As it does so, it cools by expansion and its relative humidity rises. Suppose the parcel is sufficiently moist that it reaches saturation before arriving at the top of the mountain. That means its LCL, its cloud base, would be somewhere along the windward side. A cloud would form, its base at the LCL and extending to the mountaintop.

If you're at the top, it's a fog because the cloud base is at your ground. What happens to the parcel once it reaches the top? That's an interesting question. What will it do? Before we address that question, we need to look at this process on our temperature/height diagram. So let's start with a subsaturated air parcel. We have a temperature, we have a dew point, and we know now that as we lift the parcel, their paths will cross. So let's lift it to the LCL and we're now at cloud base. Temperature and dew point are both lower than when they started, but now they're the same as well. Now let's lift it even farther, up the moist adiabat to the mountaintop. We're going up the moist adiabat because now we're saturated.

We can compare the parcel's temperature to what it would have been if it had never become saturated over that same ascent. That parcel's temperature is indicated by the hollow yellow circle. Condensation warming has partially offset expansion cooling, and that's why our saturated parcel is a fair bit warmer than our subsaturated parcel would have been. So here's our cloud, extending from the LCL up to the mountaintop. We've now seen 2 different ways of looking at this cloud. So what's next? What happens next at this point depends on whether the parcel is more or less dense than the environment air at mountaintop height. If the parcel is positively buoyant, it continues to rise and creates a deep convective cloud. If the parcel is negatively buoyant, colder than its surroundings, it will sink back down the leeside, leaving behind the mountain cap clouds that we saw in the movie.

Let's look at this latter situation. I've sketched an environment with a typical ELR. The environment's temperature at mountaintop is indicated by the white dot. The moist adiabatic lapse rate is indeed smaller than the environment lapse rate, at least in the lower troposphere, where there can be a lot of moisture to convert to heat. But because our parcel started subsaturated and cooled at the fast dry rate first, this parcel had no chance of becoming warmer than the environment in this particular case. Our parcel is more dense than its surrounding cloud-free air, so it's going to sink down the leeside of the mountain, but how it descends down the other side of the mountain depends on the answer to a very important question: Did the cloud precipitate or not?

This actually makes a very big difference. So now, let's consider 2 extreme scenarios. I'll call them all cloud and no rain, and all rain and no cloud. First, all cloud and no rain. If the parcel loses no condensation on the way up, it goes down exactly the same path on our temperature/height diagram that it went up. The reason is because as the parcel starts sliding down the other side, parcel descent causes compression warming. That increases the temperature. That makes the vapor capacity rise. To keep from becoming subsaturated, condensation will re-evaporate to vapor. So we go down the moist adiabat, the same moist adiabat we went up on, until we run out of condensation, that is. That's the LCL on the other side of the mountain, cloud base on the way down. From there on down to the surface, we warm up at the dry adiabatic lapse rate.

And if we return to the same elevation on the leeside, the same elevation as we started with on the windward side, we see we regain precisely the same temperature and dew point we started with. It's as if we never scaled the mountain at all. But what if we lost some condensation to rain or snow? What if we lost all of it? This scenario is all rain and no cloud on our temperature/height diagram. The air parcel has to descend dry adiabatically, all the way from the mountaintop. Why? Because there's no condensation to oppose compression on the way down. All the condensation was left behind on the mountain slope. So note the temperature ends up warmer than it started. It could even be a lot warmer, and the dew point increases 2°C per kilometer of descent, but it ends up lower than its original value as well.

Since the dew point tells us the vapor supply, we know our vapor content is lower. Of course, we lost some vapor to rain on the mountain, so the relative humidity of the air on the leeside, higher temperature, lower dew point, higher vapor capacity, lower vapor supply, the relative humidity of that air can be very low. So why do we often find deserts on the leeside of mountains? Because mountains can steal moisture from air by lifting air to saturation, by helping them rain, and what clouds leave behind is not available to the air on the way back down.

Now, we looked at 2 extreme scenarios, and reality is somewhere in between. Still, mountains play a very major role in determining the weather and climate of a particular place, and one reason is their ability to extract moisture from air by lifting air and making it rain. So now it's time for some instability. It's time to look at a thunderstorm sounding. What I mean by "sounding" is a vertical profile of temperature, dew point, and other properties as well, a case in which it's possible for a surface parcel to become positively buoyant on ascent, despite starting off subsaturated and struggling.

Here's our environment, temperature decreasing in the troposphere, up to a tropopause at about 12 kilometers or so above sea level. Let's take a surface parcel, a subsaturated surface parcel, temperature less than dew point. And we're experts at this now. Let's lift it to saturation, to its LCL, and we see that it's colder than its surrounding environment. It's negatively buoyant. But let's keep lifting. We have some source of sustained ascent. We have a mountain, a front, some other means of pushing air aloft, even against its own will. The now-saturated parcel follows the moist adiabat, and look as its new path is catching up with the environmental temperature.

Despite its initial disadvantage, the parcel actually manages to become warmer than its surroundings. It's reached its LFC, its level of free convection. At this point, it's the same temperature or higher than its surrounding environment, and it's convecting freely. It does not need an external push any more. The parcel is positively buoyant. We have free convection. From here, the sky is the limit, or the tropopause, which usually comes first. The rising parcel will eventually exhaust its supply of water vapor, its rocket fuel. Its gas tank is going to go empty. It also approaches the stratosphere, a region of high stability. In any case, eventually, it becomes colder than its surroundings again.

At the equilibrium level, marked on the figure as "EQL," at the equilibrium level, the parcel is the same temperature as its surroundings again. Its positive buoyancy is gone. Inertia may carry it upward a little farther, but this parcel isn't going to get much farther away from Earth. Another term for EQL is "cloud top." It's our first guess at cloud top. Actually, if you look closely, I've drawn cloud top at a level a little above the EQL since, as I just said, rising air can overshoot it a little bit.

Let's check out temperatures in the cloud, relative to the cloud's surrounding environment. The main part of the cloud is warmer than its surroundings, and positively buoyant. The lower section, between the LCL and the LFC, between the lifting condensation level and the level of free convection, is colder. Air had to be pushed to get there, and it has to be pushed to stay there as well. The very top part, the overshooting top, above the EQL, is also colder than its surroundings. The strongest updrafts lie beneath the highest overshoots, and the overshoots collapse as the updrafts weaken beneath them.

We made a cloud because we lifted air to saturation and beyond. As the relative humidity reached 100%, though, the excess vapor needed to condense and it needs surfaces to condense on. Those are our condensation nuclei—sand, salt, soot, dirt, actually, pretty common, lots of the stuff floating around, even in so-called "clean air." But remember, even if our cloud is warmer than its surroundings, it's still getting colder with height as well. Somewhere in the middle, usually about 3 to 5 kilometers above the sea level, the cloud is at the freezing level. The so-called "freezing level," I should say, because cloud drops there may not freeze, at least right away, owing to a lack of ice nuclei, as we discussed before.

Remember, liquid water is very finicky about what it'll freeze on. Ice wants to form on something that already looks like ice, and there's a shortage of those, especially in the cloud layer where the temperatures are between freezing down to $-15°C$ or $-5°F$. This is the zone of supercool liquid. The temperature is below freezing, but there are no suitable surfaces to freeze on, until maybe an airplane flies through. That's the mother of all ice nuclei.

Let's consider the region of positive buoyancy. This cloud is only as deep as it is because that air had an LFC and it made it there. Now, note all the distance that the parcel was warmer than its environment, from the LFC all the way up to the EQL. That area represents potential energy that can be used by the parcel to rise, and possibly rise very quickly. We call it "CAPE," C-A-P-E, convective available potential energy. It's another one of those self-defining terms. It's energy potentially available to convection to clouds. We also call it "positive area," but before the CAPE can be tapped, negative buoyancy has to be overcome as well. That's the negative area on our sounding. We call this "convective inhibition" or "CIN," and we pronounce it "sin." CIN is bad. You may have heard that in a somewhat different context.

Our sounding has an LFC for surface air, and that air can become positively buoyant if lifted high enough, but that doesn't mean it has the means to get there. Just because you have an LFC doesn't mean you'll get storms. That's a serious forecasting problem. Will there be enough lift? Will the lift be strong enough? Will it last long enough? Will it overcome the CIN? Imagine Stick Man is walking along the beach and he kicks a moist but subsaturated air parcel, sending it skyward. But let's say the kick wasn't very hard, and the parcel doesn't even reach its LCL before returning to the beach, never having made a cloud.

Well, let's consider a different scenario. Let's say this time, he gives the parcel a somewhat harder kick, and this time, the parcel reaches its LCL, forming a shallow little cloud. But this air never becomes warmer than its surroundings. It never even drizzles. It just comes back down the same way that it went up. But now, let's say Stick Man had an extra biscuit for breakfast and gives the parcel a crushing kick. The parcel not only clears its LCL, but also manages to snag its LFC. At this point, the air is positively buoyant and it will rise on its own. Fueled by condensation warming, this parcel will rise quickly towards its EQL. It'll probably even overshoot a little bit, before settling back down.

It's found a new home and never returns to the beach, unless it becomes part of the cloud's downdraft, that is. Air in clouds could also rush downward at high speeds as well. We'll look at thunderstorms in more detail in a future lecture, but for now, let's return to our old friend, the sea breeze. The sea breeze initiates lots of thunderstorms, and we now have the tools to understand why.

It's a sunny day. As morning wears into afternoon, the land surface is heating up, and the thickness of the layer between 2 isobars is increasing. The pressure difference between land and sea is established above the surface. Air is rising over the land, spreading towards the sea. The movement of air out of the column over land is reducing the surface pressure. That air is crowding into the column over the sea, increasing the surface pressure offshore. We have now established a pressure difference at the surface between sea and land, and the cool sea air starts moving inland. This circulation was generated by temperature differences, the ascent of air over the warm land and sinking over the cool sea. And these motions are helping to reduce that temperature difference, but the marine air remains cooler.

All of this we have previously established. Now, let's take the next step and consider what happens to the cool sea air as it pushed inland. As you can imagine, it will encounter the warmer air that already resides over the land. But here's a very important point: Fluids with different densities do not want to mix. Fluids with different densities do not want to mix. When the warmer air over the land and the cooler air from the sea collide, they do not mix and a front is formed. Those air masses not only have different temperatures, but also different densities, and we've made a front that we call the "sea breeze front."

The sea air is going to burrow beneath or underrun the less dense air over land, forcing the land air to rise over the sea breeze front. If you live near the coast, you've probably experienced sea breeze frontal passages many times. You'll remember that the temperature dropped and the wind direction changed, and probably also increased in speed as well. If the land air is sufficiently moist and if the frontal lifting is strong and deep enough, we can even lift that land air to its LCL, creating a cloud above the sea breeze front. At least at first, this cloud will probably be shallow, non-precipitating, and not positively buoyant.

However, if the land air has an LFC, and that LFC can be reached, the sea breeze lifting can provoke deep convection, leading to strong winds and heavy rain. These storms may also be thunderstorms, which are storms that produce lightning and thunder. Not all storms are thunderstorms. Here is a map of mean maximum daily precipitation for the lower 48 U.S. states, averaged throughout the year. It shows where the largest or most persistent rainfall is. We see the largest values at the Gulf Coast, stretching up the eastern seaboard. There are local maxima on the West Coast as well, north of San Francisco and in the mountains of California and the Pacific Northwest.

We can compare that to this map, which shows the average number of days per year of thunderstorm activity. We see that thunderstorms are very rare along the West Coast, but by far, the region of highest frequency is along the Gulf Coast, including Florida. Many of these thunderstorms were initiated by the sea breeze front. You're looking at visible satellite imagery from NOAA for a day in May during 2003. This is Florida. We see southern Florida, and sea breezes are developing all along the coast as the land heats up during the day. Clouds and then deep clouds are appearing on each coast as the sea breeze front pushes inland.

This is the image for 9:45 am eastern daylight time, but remember, meteorologists use London time, and that's why the image is stamped 13:45Z. It's 1:45 pm London standard time. At this very early time of 9:45 am, we see very little in the way of clouds over south Florida, but the land surface is heating up relative to the adjacent ocean areas, and the sea breeze is starting to build. By 2 hours later, 15:45Z, 11:45 am, the situation is much different. Clouds now appear over much of the land surface, being best developed near the southwest shoreline. The land is getting hot. The land air is rising, and some of that rising air has found its LCL.

Meanwhile, cooler denser marine air is also pushing inland, lifting land air that is moist enough to reach its LCL. That's good organized lifting, so the thickest clouds are found along the sea breeze front. The front is best developed on the southwest coast at this time. If we look closer at the clouds overland, you'll notice that they tend to form thin parallel lines and look like pearls on a string. We often call these clouds "streets" or "roll" clouds. They result from strong uneven heating of the land, and usually appear by the early afternoon. By 17:45Z, or 1:45 pm, local time, thunderstorms have erupted over south Florida. These storms formed as the marine air pushed inland, lifted the moist air over the land. We not only reached the LCL now, but also the LFC.

Notice the roll clouds over central Florida are also now better developed at this time, and also the roll clouds are everywhere, except over Lake Okeechobee. The lake is not as strongly heated as the land surfaces are. So we saw roll clouds over Florida form as the Sun heated over the land surface. As the land heats up, the relative humidity decreases because typically, the vapor supply is not changing. That would make it seem harder to create a cloud by lifting. The air is actually getting relatively dryer and you have to lift it farther to bring it to saturation. Despite that, it's actually easier to make clouds in the afternoon, and here's why.

Say this is our environmental profile, our environmental temperature profile at some time in the morning. Notice we're only looking at the lowest 1.5 kilometers, or 1 mile, above the surface. Let's make an air parcel of surface air, subsaturated so we see the temperature is higher than the dew point. And by lifting this parcel dry adiabatically, I see I can bring it to saturation with a vertical push of about 600 meters or so. That's about 2000 feet. I may not even be able to get this parcel this far because I may not have a source of lifting, but suppose I do. Further lifting of my now saturated parcel will follow the moist adiabat, but this parcel is never able to become warmer than its surroundings. It's always negatively buoyant, so chances are our morning weather is clear sky and no clouds. But as we move into afternoon, solar heating is making the land surface a whole lot hotter.

The land surface probably has small thermal inertia, and it can't stop itself from warming. Air is a lousy conductor, so upward transport of this energy is pretty slow. As a result, the environmental lapse rate is getting larger near the surface. Meanwhile, the surface dew point probably didn't change all that much because nothing has changed the vapor supply. The temperature and the vapor capacity have gone up, the dew point and the vapor supply has not, so the air now has an even lower relative humidity than it had earlier in the morning.

If we want to make a cloud out of this surface air, we have to lift it even farther now. Note the LCL is now located about 400 meters farther above the ground. But also note there's much less resistance to lifting now. In fact, note that the environmental lapse rate has actually gotten larger than the dry adiabatic rate over the lowest 800 meters. This atmospheric layer near the ground has become absolutely unstable. Even a little push of this parcel upwards will permit it to continue rising, even though it's not saturated. Even Stick Man could lift this parcel. But the parcel doesn't remain positively buoyant very long, and actually, it will probably not get too much above its LCL, so what we'll end up with is these tiny little clouds, which we see in the afternoon from the beach as our pearls on a string.

Let's summarize. Temperature in the standard atmosphere decreases with height in the troposphere. That's the environmental lapse rate, one of our 4 lapse rates. Let's make an air parcel and lift it. As long as it has moisture, it will saturate on ascent because the air's temperature drops faster than its dew point. Once saturated, it will cool more slowly at the moist adiabatic lapse rate, and it might become positively buoyant, given sufficient lift. But what goes up must come down, if only somewhere else, and some other time.

How air comes down, say on the back side of a mountain, depends on how it went up. Did it keep its condensation? Did it lose some or all of it to rain or snow? The answer to that question may determine whether a desert exists on the leeside or not. Some days, there's no way to get a storm. You lack lift. You don't have the moisture. You don't have an LFC. Other days, you do have a LFC, but not enough lift to get air there. On those days, you won't be able to get to the CAPE, the convective available potential energy, and that's a big forecast problem. CAPE is fuel for storms, positive buoyancy that drives air upward. We'll see how it is used in organized storms, like the squall line, and the fearsome supercell in future lectures.

But now, we need to turn our attention to horizontal winds—what drives them, what determines their direction and intensity? We put our back to the winds in the next lecture.

Lecture Eleven
Whence and Whither the Wind, Part 1

Scope:

In this lecture, we examine forces that work counter to nature's tendency to push air from high to low pressure, from colder places (the north) to warmer places (the south). The 4 principal forces that determine where, when, and how quickly the horizontal wind blows are pressure gradient force (PGF), the Coriolis force, frictional force, and a centripetal or centrifugal force. In this lecture, we'll discuss the first 2 of these forces and learn how they participate in determining how and where the wind blows.

Outline

I. The 4 principal forces that determine where, when, and how quickly the horizontal wind blows are pressure gradient force (PGF), the Coriolis force, frictional force, and a centripetal or centrifugal force.

   A. A pressure gradient is a pressure difference divided by a distance. Pressure differences drive winds, but pressure gradients determine wind speed.

   B. The Coriolis force owes its existence to the Earth's rotation. This force acts to the right in the northern hemisphere and to the left in the southern hemisphere; it vanishes at the equator.

II. During World War I, the Germans deployed a long-range cannon, the Paris Gun, directly east of Paris and fired shells at the city. According to some accounts, the Germans failed to take the Coriolis effect into account, and many of their shells fell to the north of Paris.

   A. A rocket fired at the North Pole travels straight, but from our point of view, it seems to curve to the right because what we call north is not fixed in time.

   B. Our compass directions also vary in space due to the fact that the Earth is a sphere. A rocket fired to the west cannot travel without a change of latitude because the Earth is curved, and latitude circles are curved, as well. We explain this apparent deflection with another convenient untruth, the curvature force.

C. In the incident from World War II, the curvature force was working in the opposite direction to Coriolis. Coriolis tried to deflect the shells to the north, while from our point of view, curvature pushed them to the south.

D. In fact, based on the reported shell velocity, the curvature deflection would exceed the Coriolis deflection, and the Paris Gun shells should have missed Paris to the south.

III. Why does the Coriolis force act to the left, following the motion in the southern hemisphere? And why does this force vanish at the equator?

A. Seen from above the North Pole, the northern hemisphere turns counterclockwise. Seen from above the South Pole, the southern hemisphere turns clockwise. The Coriolis force acts to the left, following the motion in the southern hemisphere.

B. To answer the question of why the Coriolis force vanishes at the equator, let's consider vectors. A vector is an arrow that represents a path and can be broken down into its components.

C. Think of the spin axis of the Earth as a vector, whose orientation is determined by the right-hand rule. From the North Pole, the Earth's spin axis is straight up and down; it's in the local vertical.

D. As we move toward the equator, the vertical component of the spin axis becomes smaller.

E. The Earth is still spinning, but less of that spin is working to rotate the coordinate system at that latitude. The Coriolis force represents the spin vector in the local vertical. At the equator, the vertical component of the Earth's spin axis is zero.

IV. The wind that results when both pressure gradient and Coriolis force are active (active against each other) is called the geostrophic wind. This is the wind balance that is created because the Earth turns.

A. Consider a pressure difference between 2 isobars and an air parcel. The parcel experiences a pressure gradient force directed toward low pressure. Once it begins to move, the Coriolis force appears, directed to the right. The combination of these 2 forces causes the parcel to deviate to the right of its original path.

B. The Coriolis force continues to turn the parcel to the right until the pressure gradient force and the Coriolis force are locked in opposition. In this situation, the wind is blowing parallel to the isobars with low pressure to the left. This is geostrophic balance.

C. The wind is not moving toward low pressure, the direction the pressure gradient force is trying to push it. The wind is not turning to the right, following the motion, as the Coriolis force is trying to push it. Instead, it's a straight-line wind.

D. We cannot disturb the geostrophic balance by pushing large-scale wind harder. Ultimately, the pressure gradient force determines the wind speed, but the Coriolis force is proportional to wind speed.

E. At sea level, if the wind is geostrophic, we usually know that it's blowing with low pressure to the left. Its speed will increase as the isobar spacing on a sea level pressure chart becomes smaller.

V. The geostrophic wind blows parallel to, not across, isobars. But isobars are not always straight, and to make air flow parallel to curving isobars, we need to add another force to the mix. Sometimes, we can make air blow across isobars from high to low pressure, but that requires another force.

Suggested Readings:

There is no suggested reading for this lecture.

Questions to Consider:

1. If the Earth did not rotate on its axis, would we still have winds? If so, what would the average wind direction be in Los Angeles?

2. It has been claimed that, in the northern hemisphere, the right-hand rails of railroad train tracks wear out faster than the left-hand rails. Why? (Presume only one direction of motion is permitted on these tracks.)

Lecture Eleven—Transcript
Whence and Whither the Wind, Part 1

Welcome back to Meteorology, where we continue our survey of the wonders of the weather. Early on, I made this statement: "Temperature differences make pressure differences, and pressure differences drive winds." We saw this was true in the case of the sea and land breezes, circulations that are established owing to temperature differences across a coastline. There are a lot of reasons why one place might be warmer than another. It might receive more direct solar radiation, owing to Earth's sphericity and tilt. It might receive more infrared radiation from the atmosphere, owing to a greater concentration of greenhouse gases, especially water vapor, and the presence of low clouds. It might absorb more of the radiation it does receive because of the characteristics of its surface. It might not be able to resist the radiation it does absorb, owing to a low thermal inertia. It might experience more subsidence, descending motion, increasing the temperature by compression warming. More water vapor may be condensing there, liberating latent heat.

Here's another statement that I've made that we've already seen an example of: "Nature wants to push air from high to low pressure." Nature abhors extremes, excesses, and imbalances, including the imbalance of having more atmospheric mass in some places than others. Where there's more mass piled up, the pressure is higher, at least at the surface, and Nature wants to push that air away, towards a place of relative deficit, having a lower pressure. These 2 combinations help to explain the sea breeze, as well as the land breeze, examples of what we called "thermally direct" circulations. But both are also part of a circulation that has surface winds, blowing from a colder to a warmer place. The cool wind could cause a front as it pushed inland because air with different densities do not want to mix. Lifting at a front can cause clouds, if not also storms, but these breezes weren't always cool. We saw that as well, and that was when air had to flow downslope and experience compression, as in the case of the Santa Ana winds.

Ultimately, the most important temperature difference is between the equator and the pole, established because of the Earth's spherical shape. If we applied the sea breeze model to the Northern Hemisphere, we would see it predicting a surface wind, blowing from the cold north to the warm south, from the pole towards the equator, but sustained surface winds don't blow north and south in very many places. In the middle latitudes, roughly between 30 and 60 degrees north, the prevailing wind is from west to east, at least in places not directly influenced by topography. In the tropics, southward of 30 degrees north, the prevailing winds blow from east towards west, so I need to qualify my statement. Now I will say, in the absence of other forces, Nature will push mass from high to low pressure. In the absence of other forces, Nature will push mass from high to low pressure. That's quite a concession.

The simple fact is, except in simple situations like the sea breeze, other forces aren't absent. Nature wants to move surface air from north to south, but these other forces usually serve as spoilers, interfering with Nature's plans and raising her ire. In this lecture, we start our examination of those other forces, and how they participate in determining how and where the wind blows. The 4 principal forces that determine where, when, and how quickly the horizontal wind blows are these: First, pressure gradient force, PGF. Second, the Coriolis force, due to Earth's rotation. Third, the frictional force, and finally, a centripetal or centrifugal force, depending on your point of view.

Two of these 4 forces, pressure gradient force and friction, are real forces. Coriolis is an apparent force, and the centripetal/centrifugal is a little bit of both. Along the way, I'll mention a fifth force, a curvature force, owing to Earth's sphericity. That's another apparent force, and one we usually neglect, except with very fast motions, such as rockets. You may be wondering why are we even bothering with forces that aren't real? We'll see.

So first, pressure gradient force. We have seen that while pressure is a very important force, pressure differences held the key to understanding the barometer and the water-filled straw. The next step is to recognize the role of pressure gradients. A pressure gradient is a pressure difference divided by a distance. The pressure gradient is large when the pressure difference is large, or the distance between the 2 are small. Pressure differences drive winds, but pressure gradients determine wind speed, the velocity, and the ferocity of the horizontal wind.

Average sea level pressure is about 1000 millibars. What would a large pressure difference be? How about 10%? So 10% would be 100 millibars. Let's picture a 100-millibar pressure difference between 2 places. What kind of wind would this pressure difference produce? It depends on the distance. If the 2 locations were separated by the entire hemisphere, the wind would be very weak. The pressure difference is large, but the distance is enormous, and the pressure gradient is small. That same pressure difference of 100 millibars over a distance of only a mile or 2 would generate the enormous winds of the EF5 tornado, the strongest winds on Earth.

Our second force is the Coriolis force, or Coriolis effect, which exists owing to Earth's rotation. It's named for Gaspard-Gustave de Coriolis, the 19th-century French scientist who first explained the effect. The Coriolis force acts to the right, following the motion in the Northern Hemisphere, and to the left, following the motion in the Southern Hemisphere. It vanishes at the equator. It's an apparent force, one that exists owing to our point of view, and yet, we can use this force to explain why if we shoot a rocket straight towards the North Pole, we will miss. We can use it to explain why hurricanes don't form on or across the equator. We can use it to explain why Nature can't make a simple sea breeze circulation that spans the hemisphere. These phenomena are real, but our explanation is self-serving.

Let me make an analogy with the rising and setting Sun. It's morning and it's bright. Flowers open, bees start their rounds. The Sun has risen in the east and temperatures are rising. The Sun will cross the sky, ultimately setting in the west. The nighttime will be dark. Flowers will close and bees will sleep, and temperatures will drop. All of these phenomena are real, but if we explain that all this occurs because the Sun is revolving around us, our explanation is wrong. And yet it's a very tempting point of view. The Earth actually rotates very quickly. Its spin velocity at the equator exceeds 1000 mph, and yet we cannot feel it. It's much easier to believe that the Sun is moving and we are fixed in space. It conforms with our observations and more importantly, it simplifies our explanations.

We construct a frame of reference in which we are at the center of it all. Everything revolves around us. In theory, our explanations are wrong, but in practice, they're convenient and they work. The Coriolis force is another convenient untruth. Let's pretend for now that the Earth is a flat rotating disk, and let's look down on the North Pole. Chicago and New Orleans are roughly at the same line of longitude. They're roughly north and south of each other. Suppose the Earth were not rotating and I fired a rocket from New Orleans towards the pole. The rocket would fly straight and arrive at the pole just fine, having passed over Chicago on the way.

But now consider Earth's rotation. After a few hours, Chicago and New Orleans will have shifted counterclockwise, at least seen from space because the Earth rotates counterclockwise, looking down on the North Pole. On the Earth, I'm still in New Orleans, and looking north, I still see Chicago, but as seen from space, north has changed. Newton's first law of motion states this: An object placed into motion continues moving in a straight line at constant speed unless other forces are acting. We launched a rocket. No additional forces are acting. The rocket flies straight.

But that's not what we see. From our vantage point, we see the rocket has turned and it's turned to the right, following its motion. We fired the rocket straight north, but it didn't stay going north. Almost immediately, it started curving to the east. And in fact, if I were depending on this rocket keeping a northward track, I'm in trouble. On the rotating Earth, a rocket fired directly towards the pole will never get there. The rocket didn't stay moving north, but the rocket didn't turn—north did. We see the rocket has turned, but the rocket didn't turn—we did. But the bottom line is this: However we choose to explain the rocket's motion, it does not reach the pole.

If we can't do it with rockets, Nature can't do it with air. Nature can't make the wind blow directly from the pole to the equator, or the equator to the pole. But it remains that, from our point of view, the rocket turned. Again, Newton's law of inertia implies if an object deviates from straight line motion, there must have been a force. In truth, there was no such force because the rocket didn't really turn, so we invent a force, an apparent force, to explain what we see, and the Coriolis force is our explanation for why the rocket turned from our point of view. The phenomenon is real. The explanation is self-serving.

Of course, we can reach the pole with a rocket on the rotating Earth. We can do it by adjusting the rocket's motion along the way, or by adjusting our aim to account for the Coriolis effect. We'll just shoot away from the pole, and let the Coriolis effect turn it into the pole. But in any event, it's important to recognize that the Earth did not turn beneath the rocket.

Speaking of rockets, there's a popular story regarding Big Bertha, a long-range cannon deployed by the Germans against the French during World War I. The Germans placed their cannon over 100 kilometers directly east of Paris and fired shells at the City of Light, startling the French, as the bombs fell seemingly from nowhere. But some of the stories claim that the Germans forgot to take the Coriolis effect into account, so their shells kept missing Paris, falling to the north, which would be to the right, following the motion. Some claim the shells curved so far north, they didn't even land in France. The true story is both more prosaic and more curious.

First, the prosaic part. First of all, the gun that the Germans trained on Paris was not the Big Bertha that they used elsewhere, earlier in the war. Actually, it was a further advanced weapon called the "Paris Gun." And according to reports, this supergun was able to push a shell a distance of 120 kilometers in about 3 minutes, passing the tropopause on the way and becoming the first human-made object to reach the stratosphere. The shell covered this relatively short distance so quickly that the Coriolis deflection on the rocket would only have been a kilometer or so. Given the limited accuracy of the gun and the size of the target, it's not clear that the Coriolis deflection would even have been noticed.

But now for the curious part. Remember that, from our point of view, Coriolis explained a curving motion that we saw that didn't really happen. The rocket went straight. We saw it curve to the right because what we call north is not fixed in time, owing to Earth's rotation. The curvature force that I mentioned briefly earlier is similar to this, in the sense that our compass directions, north, south, west, and east, also vary with location on the Earth, due to the fact that the Earth is a sphere. Another way of saying that—they vary in space.

Consider any location on the Earth, other than the poles, and consider the east/west direction. That's the direction in which latitude is unchanged. If we fire a rocket to the west, we expect it to continue moving straight west without change of latitude, but it cannot do so because in reality, because the Earth is curved, latitude circles are curved as well. The rocket in this case does fly straight, but in this case, we see it curve to the south. A rocket fired to the east also appears to curve to the south. But the rocket fired straight west or east can't keep going in those directions, and it's because west and east aren't straight.

We explain this apparent deflection with another convenient untruth, the curvature force. For slower motions, such as the winds, we can safely neglect this force, but for high-speed rockets and projectiles like the Paris Gun shell, it's actually very important. Note that in our example, the curvature force is working in the opposite direction to Coriolis. Coriolis was trying to deflect the shell to the north, following the motion. Curvature is pushing it to the south from our point of view. In fact, based on the reported shell velocity of 40 kilometers per minute, which is about 700 meters per second, or 1500 miles per hour, the curvature deflection would exceed the Coriolis deflection, and the Paris Gun shell actually should have missed Paris to the south.

Two more questions. Why does the Coriolis force act to the left following the motion in the Southern Hemisphere? And why does the Coriolis force vanish at the equator? The first one is easy. Take a globe and look down on the North Pole and give it a counterclockwise spin because the Earth is spinning counterclockwise as seen from above. But now, turn it over and look down at the South Pole. The Earth is turning the opposite way. Thus as the Northern Hemisphere is turning counterclockwise as seen above the North Pole, the Southern Hemisphere is turning clockwise as looked above the South Pole. So the Coriolis force is acting to the left, following the motion in the Southern Hemisphere.

Why does Coriolis force vanish at the equator? This one is a little more complex. The reason is again, because the Earth is a sphere, and it's not a flat board or anything else either, as I used in that previous description. First, let's consider something about vectors. A vector is an arrow, generally representing a path. Any vector can be broken down into its components. Suppose this vector represents motion towards the northeast. Part of that motion is towards the north, and part of that motion is towards the east. Combined, it produces the total motion, which is towards the northeast.

Let's consider our spherical rotating Earth. As seen from above, the North Pole, as I said, the Earth is spinning counterclockwise, about a spin axis that is oriented vertically. We know that this axis also tends to be tilted relative to the Sun, but that doesn't matter in this situation. We can treat the spin axis of the Earth as a vector. Its orientation is determined by what we call the "right-hand rule." I take my right hand and I curl my fingers in the direction of the rotation, in this case, counterclockwise. And then I look to see where my thumb is pointing. For a counterclockwise curl, my thumb is pointing up, so the Earth's spin axis is a vector that points straight up. If you're at the North Pole, the spin axis is in the local vertical. It's pointing straight up from where you are.

For the wind blowing in the horizontal, though, the only component of the spin vector that actually matters to us is what's in the local vertical, not what is relative to the Earth's North Pole because that's the spin that rotates our coordinate system, including north. We refer to the magnitude or length of the vertical component of this vector as "f" and f is a maximum at the North Pole. The vector has only one component there, pointing in the local vertical. But as we move towards the equator, the spin axis direction is the same, but it's no longer in the local vertical any more. So the important vertical component of the spin axis is becoming smaller and smaller.

The Earth is still spinning, but less of that spin is working to rotate the coordinate system at that latitude. It is the component of the spin vector in the local vertical that the Coriolis force represents. At the equator, the Earth's spin axis is actually horizontal relative to the ground. The vertical component of the spin axis is zero, so the Coriolis force vanishes at the equator. The Earth spins, but north does not.

For the large-scale wind that wants or needs to travel a considerable distance at moderate speed, the 2 most important forces are pressure gradient force and Coriolis. I will define the large-scale wind as a wind in which Earth's rotation, expressed by the Coriolis force, is important. The pressure gradient force is the driving force of the horizontal wind. The Coriolis force is always present on a rotating Earth, but becomes important as the velocity, length, and timescales associated with the motion become larger. The sea breeze circulation, spanning a few dozen kilometers in a few hours, is too local in nature to be profoundly affected by the Earth's rotation, but the circulation Nature tries to establish in reaction to the equator to pole temperature difference—that spans a hemisphere and Coriolis plays a powerful frustrating role.

The wind that results when both pressure gradient and Coriolis force are both active, and active against each other, is called the "geostrophic wind." "Geo" means Earth, "stroph" to turn. We have already seen that in the word "troposphere." This is the wind balance that is created because the Earth turns. So let's see how the wind comes into geostrophic balance. We call this "geostrophic adjustment."

Let's start with a pressure difference between 2 isobars, drawn in a horizontal plane. We can take north at the top of the figure, if we so choose. And let's create an air parcel, but hold onto it for now. The parcel is experiencing a pressure gradient force directed towards low pressure, but it's not moving because we got our mitts on it. But once we let go, once the air parcel is free to move, it will start moving, and it'll start moving towards low pressure. But once in motion, the Coriolis force now appears, directed to the right, following the motion. The combination of these 2 forces will cause the parcel path to start deviating to the right of its original path, so that it is no longer moving directly towards low pressure.

The Coriolis force is always directed to the right in the Northern Hemisphere, so as the parcel starts turning because the Coriolis force is tugging on it, the Coriolis force keeps tugging rightward. So the turning continues, and the turning continues, and it will continue until the pressure gradient force and the Coriolis force finally lock into opposition. And in this situation, the final state of the wind is where the wind is blowing parallel to the isobars with low pressure to the left. We call this "geostrophic balance."

Note these facts about the geostrophic wind. It blows parallel to isobars, with low pressure to the left in the Northern Hemisphere. The wind is not moving towards low pressure, as the pressure gradient force wants. The wind is not turning to the right, following the motion, as the Coriolis force is trying to do. Instead, it's a straight line wind. Pressure gradient force and Coriolis force have come into a stalemate. What we call the "Buys Ballot law," named after a Dutch meteorologist, goes like this: In the Northern Hemisphere, stand with the wind at your back, and then low pressure is towards your left. Actually, and we'll understand this better later, it's not straight left because chances are, the wind at your back is not purely in geostrophic balance, but we're not done examining our other forces yet.

Why does this balance come about? Could we disturb the geostrophic balance by pushing the wind harder? If we push the wind hard enough, could we get the wind to move towards low pressure? The answer is no, not for the large-scale wind. Ultimately, the pressure gradient force determines the wind speed. Weak pressure gradients mean weak winds, but the Coriolis force itself is proportional to the wind speed. An object at rest on the Earth's surface does not care that the Earth is rotating. There is no motion to be affected. But the faster the object's motion over the rotating Earth, the more profound the influence of the coordinate system rotation is, and that's what the Coriolis force really is.

This means if we increase the pressure gradient so the wind speed increases, the Coriolis force will intensify in equal measure, so the wind remains flowing parallel to the isobars in geostrophic balance, just a whole lot faster. On sea level pressure charts, which we'll start examining pretty soon, we draw isobars, lines of constant pressure, at a regular interval, typically 4 millibars. The pressure gradient then is determined by the spacing of the isobars. The more closely they're spaced, the larger the pressure gradient force.

Consider, as an example, 2 relatively widely spaced isobars between values of 996 and 1000 millibars. The pressure gradient force is directed towards low pressure, but it's not especially strong. As a result, the wind speed is not especially fast, which means the Coriolis effect is not particularly large. The geostrophic wind is blowing with low pressure to the left, but it's not blowing very quickly. Now, let's pack the isobars more tightly. The pressure gradient force is now larger, driving a faster wind. Faster winds across the Earth's surface lead to a more substantial Coriolis effect. The 2 forces still come into balance, but the wind is stronger and note now it's still blowing with lower pressure to the left.

This means we don't actually need to draw the winds on a sea level pressure chart to have a very good guess regarding either wind direction or speed. If the wind is geostrophic, it's blowing with low pressure to the left. If the wind is geostrophic, wind speed increases as the isobar spacing becomes smaller. Let's do an example. What is the wind direction and speed in this case? Keep in mind that low pressure in this case is towards the northwest. Remember, we name winds by the direction they're blowing from. This wind is blowing from the southwest towards the northeast, so we call it a southwesterly wind in the Northern Hemisphere. The wind speed is moderate, and that's determined by the isobar spacing.

Let's do another case. What is the wind speed and direction in this situation? Now the low pressure is to the southeast. This is a northeasterly wind, blowing with the same speed as my previous example because the isobar spacing is the same. Decreasing the isobar spacing implies a faster wind if the wind is in geostrophic balance. Therefore we don't need to sketch the winds. We just need to examine the isobars, how they're oriented and how they're spaced.

Since the geostrophic wind blows parallel to and not across isobars, we might find ourselves tempted to treat isobars as material surfaces, as if they were channels or walls. That might help us appreciate why the wind speed increases as the isobars become more tightly packed. Our intuition tells us if we force a fluid like air or water to pass through a narrower channel, its velocity will increase. That is a helpful example. However, we need to keep in mind that isobars are not material surfaces, and there are important situations in which air can cross them and thus flow from high to low pressure. The local sea breeze was one such example.

Let's summarize. We're considering the winds and the direction and speed at which they blow. There are 4 fundamental forces that determine the horizontal wind. We've seen 2 so far: pressure gradient force and Coriolis. Pressure gradient force is directed from high to low pressure, and in the absence of other forces, Nature moves mass from high to low pressure. The Coriolis force acts to the right following the motion in the Northern Hemisphere, to the left following the motion in the Southern Hemisphere, and is zero at the equator. The combination of these 2 forces, pressure gradient and Coriolis—geostrophic balance.

Geostrophic balance represents straight line flow, parallel to isobars, blowing with low pressure to the left, at least in the Northern Hemisphere, and the speed of the geostrophic wind determined by isobar spacing. This balance applies to the large-scale wind, not the sea breeze, but rather to air flowing over long distances and blowing for days, weeks, and longer. But the geostrophic balance has already revealed to us why the equator to pole temperature difference doesn't result in a hemispheric sea breeze, blowing along the surface from pole to equator. Earth's rotation, through the Coriolis force, has spoiled this.

But isobars are not always straight, and to make air flow parallel to curving isobars, we need to add another force to the mix. And sometimes we can make air blow across isobars from high to low pressure, even for the large-scale wind. To make the air do that, we need to add still another force to the mix. We have 2 more fundamental forces to consider, and we'll do that in the next lecture.

Lecture Twelve
Whence and Whither the Wind, Part 2

Scope:

We've seen 2 of the 4 principal forces that determine when, where, and how quickly the horizontal winds blow: pressure gradient force and the Coriolis force. In this lecture, we will encounter the other 2 forces: friction and the centripetal or centrifugal force, depending on one's point of view. Friction helps the large-scale wind move toward low pressure. Centripetal or centrifugal force comes into play when air moves in circular paths.

Outline

I. Let's begin with the third force, friction.

    A. The Earth's rotation, through its proxy, the Coriolis force, turns large-scale winds so that they blow perpendicular to the pressure gradient, with low pressure to the left in the northern hemisphere.

    B. Friction, acting near the Earth's surface to reduce wind speeds, disturbs the geostrophic balance so that the large-scale wind will blow toward low pressure.

    C. Friction weakens the Coriolis force.

    D. On a diagram, we see the wind blowing from low pressure to the left and Coriolis acting to the right of the motion. Friction opposes the motion, causing it to slow. The pressure gradient force is then able to pull the air toward low pressure, at least at a shallow angle.

    E. The result is a 3-way balance of forces, in which air can move across isobars, toward lower pressure.

II. Now, let's consider a different 3-force balance.

    A. Wind tends to turn counterclockwise around large-scale regions of low pressure in the northern hemisphere and clockwise around highs. These lows and highs are called cyclones and anticyclones.

    B. Note in our figure that the air flow is parallel to isobars of pressure. This means that friction is absent.

C. The winds' change of direction is usually explained as a centripetal or centrifugal force, depending on one's point of view. The combination of pressure gradient force, Coriolis, and centripetal or centrifugal forces is called gradient wind balance.

D. Centrifugal and centripetal forces exist when there is spin. Centrifugal force is directed outward from the center of spin. Centripetal force is directed inward toward the center of spin.

E. Note that only centripetal force is real.

F. Starting at geostrophic balance, the centripetal force acts either in the direction of the pressure gradient force, guiding the air around the low, or the Coriolis force, guiding the air around the high.

III. Where did the centripetal or centrifugal force come from?

A. Starting again with geostrophic balance, suppose an air parcel's path takes it toward a place where the isobars are curving in a counterclockwise fashion.

B. Inertia acts to push the air on a straight path, but notice that this would carry the air across the isobar toward higher pressure.

C. Notice that the pressure gradient force, which points most directly toward low pressure, opposes inertia. Part of the wind's driving force is directed against the parcel, and the air must slow down.

D. When the air slows down, the Coriolis force is reduced, allowing the pressure gradient force to change the direction of the air, except this time, it doesn't result in cross-isobar flow. Instead, the parcel continues to follow the isobars as they curve. Thus, we've made a cyclone with air curving counterclockwise around low pressure.

E. If the air parcel approaches isobars that are curving clockwise, inertia would carry it toward lower pressure. A component of the pressure gradient force is now pointing in the direction of parcel motion.

F. The parcel speeds up, increasing the Coriolis force, which makes the air bend to the right. The air remains parallel to the isobars but moves quicker than it did when the motion was purely geostrophic.

IV. Let's combine our 4 fundamental forces.

    A. Pressure gradient force and Coriolis make straight-line geostrophic flow, low pressure to the left.

    B. Centripetal gives us curvature, counterclockwise around low, clockwise around high, still parallel to isobars.

    C. Friction gives us a little cross-isobar flow into low and out of high. Notice that owing to friction, wind can cross isobars away from high and toward low pressure.

    D. In this situation, the air is converging into low pressure from below. Bringing air into low pressure at the surface creates upward vertical motion in the cyclone. Rising air can become saturated and possibly unstable, resulting in clouds and storms.

    E. Meanwhile, the surface divergence out of the high implies downward motion in the anticyclone, and we usually associate highs with stable and clear conditions.

V. We've seen 5 combinations of these 4 forces; let's look at a sixth.

    A. We saw that isobar curvature affects wind speed so that, for the same isobar curvature, as the air curves counterclockwise, the wind is subgeostrophic. The flow clockwise around highs is supergeostrophic.

    B. This is an apparent paradox because we often see strong winds around lows, and we associate highs with weak winds. In practice, isobar spacing around lows can be much smaller than around highs, and that's what leads to the faster winds.

    C. The reason we don't usually see tight pressure gradients around highs stems from the sixth combination of the forces. Pressure gradient and centrifugal force result in cyclostrophic balance. Coriolis has no role here; this is local-scale rapid spin.

    D. Spin creates low pressure, as we see when we create a vortex in a glass of water.

Suggested Reading:

There is no suggested reading for this lecture.

Questions to Consider:

1. True or False: Space shuttle astronauts in orbit about the Earth experience weightlessness because the spacecraft is so far from the Earth that its gravitational pull is negligible.

2. It is theoretically possible for large-scale wind to be able to blow clockwise around large-scale lows in the northern hemisphere. This is termed an antibaric low. For a given isobar spacing, would you expect winds around an antibaric low to be stronger or weaker than their counterparts around a normal low?

Lecture Twelve—Transcript
Whence and Whither the Wind, Part 2

Welcome back to Meteorology. We are well along in our survey of the wonders of the weather. There are 4 principal forces that determine when, where, and how quickly the horizontal winds will blow. We've seen 2 of them already. Pressure gradient force (PGF), the driving force, directed high to low pressure, the wind blowing from high towards low when acting alone. The Coriolis force, our interpretation of the effect that Earth's rotation has on the winds, acting to the right in the Northern Hemisphere, objects bending to the right when acting alone. Together, they combine to produce what we call the "geostrophic wind"; "geo" is Earth, "stroph" is to turn, the wind that exists because the Earth rotates. The geostrophic wind goes neither high to low nor does it bend to the right. This applies to the large-scale wind, which is the wind for which Earth's rotation is important, as opposed to a local-scale wind, like the sea breeze.

In this lecture, we will encounter the other 2 important forces, friction and the centripetal or centrifugal force, depending on your point of view. Friction will help the large-scale wind move towards low pressure, at least a little bit, and in those places where friction is active. Centripetal or centrifugal will be invoked when air moves in circular paths, such as around Santa Ana highs or hurricanes. Once we have our 4 forces, we'll also see various combinations of them that explain and do interesting things.

One of these will explain why the wind curves counterclockwise around our large-scale areas of low pressure. Another will explain why the large-scale wind can converge into those lows, at least at the surface. That's important because it will reinforce why we associate large-scale low pressure systems with interesting weather like storms. A third combination will confirm to us that spin makes low pressure. So let's get started. Our third force is friction.

The Earth is rotating. We have already seen that this has single-handedly prevented Nature from doing one of the things she most desperately wants to do, and that is to share the wealth of warmth, received by the tropical surface, with the frigid poles, through creating a hemispheric circulation resembling the sea breeze. The air wants to rise over the warmer equator and sink over the colder pole. Those air motions would normally lead to low surface pressure on the equator side, and higher surface pressure pole-ward,

©2010 The Teaching Company.

and a surface wind directed from pole to equator, from high to low, would result.

But then Earth's rotation, through its proxy the Coriolis force, turns the winds so they wouldn't even blow towards low pressure any more, not the large-scale wind, anyway. Instead, it made the wind blow perpendicular to the pressure gradient, with low pressure to the left in the Northern Hemisphere. That's what we call "geostrophic balance." But Nature has one last chance to tug the large-scale wind back towards low pressure, something that will disturb the geostrophic stalemate, thanks to the roughness of the Earth's surface, thanks to the Earth's surface itself, thanks to friction.

The only way we can disturb the geostrophic balance so that the large-scale wind will blow towards low pressure is to introduce a force that selectively counteracts the Coriolis effect. Friction will do that. Friction acts near the Earth's surface, and its primary role is to reduce wind speeds. Friction disturbs geostrophic balance through weakening the Coriolis force, the Coriolis force is proportional to wind speed, so as the wind is slowed, Coriolis is less effective at opposing the pressure gradient force, which isn't directly affected by friction.

Let's take a look. Let's start with geostrophic balance. I've drawn 2 isobars, low pressure to the north, higher pressure to the south. But we can turn this figure any way we want. The wind is blowing from low pressure to the left, and Coriolis is acting to the right of the motion. Friction opposes the motion, causing it to slow. Since the Coriolis force is proportional to wind speed, it is weakened. The pressure gradient force, as I said, not as directly affected, is able to now gain an advantage, pulling the air towards low pressure, at least at a shallow angle. Note that even as the air crosses the isobars towards lower pressure, Coriolis is still acting to the right. Coriolis always acts to the right in the Northern Hemisphere, no matter which way winds, objects, rockets are going, but it isn't directly opposite to the pressure gradient any more.

The result is a 3-way balance of forces, pressure gradient, Coriolis, and friction, in which air can move across isobars, towards lower pressure. Do you remember Buys Ballot's law? We can modify Buys Ballot's law to state that in the Northern Hemisphere, low pressure is roughly 15 degrees clockwise from left when the wind is at your back. Friction is typically influential only near the ground, and when, through vertical mixing, the air can come in contact with the ground and thus be slowed by contact. A fair rule of thumb is friction is very important in the lowest kilometer or so of the troposphere.

Now, let's consider a different 3-force balance, which will help us understand what we often see on weather maps. We often see the flow exhibits curvature on the larger scales, by which we mean over distances of hundreds of kilometers or miles. In particular, wind tends to turn counterclockwise around large-scale regions of lower pressure in the Northern Hemisphere, and the winds tend to curve clockwise around larger-scale highs. We call these lows and highs "cyclones" and "anticyclones" respectively.

"Cyclone" comes from the Greek term to turn in a circle, but in the Northern Hemisphere, we limit this turning to be in a counterclockwise direction, as seen from above. So "anticyclone" merely implies turning in the opposite direction as the cyclone. So anticyclonic flow is clockwise following the motion, again in the Northern Hemisphere. The names are the same in the Southern Hemisphere, but the directions are reversed. Note these facts about curving flow that I've drawn in this figure. The flow is still parallel to isobars of pressure. This means that friction is absent. Pressure gradient and Coriolis forces are still present, and they still oppose each other, and we know that these 2 forces alone produce straight line geostrophic balance. Low pressure is still to the left, and yet the winds are curving.

The winds are changing direction along the path and Newton tells us this direction change requires a force. This extra force is usually explained as a centripetal or centrifugal force, depending on your point of view. The combination of pressure gradient force, Coriolis, and centripetal or centrifugal forces is called "gradient wind balance." What are centripetal and centrifugal forces? Centrifugal and centripetal forces exist when there is spin or curvature to motion. "Centrifugal" means to flee the center. It is directed outward from the center of spin. You may have noticed on sundials, oftentimes sundials have a Latin inscription, "*Tempus fugit*," which means time flies or time flees. That contains the same root as "centrifugal."

In contrast, "centripetal" means to seek the center. It is a force directed inward towards the center of spin. These opposing forces are obverse and reverse of the very same coin, and both terms appear together in Newton's epic work, *The Principia*. So, the canonical example of centrifugal force is to use a yo-yo and twirl it. I'm going to use this rope instead, and I'm going to take this rope and I'm going to spin it over my head. Now, I do this in class a lot, and I'm sure that some of my students are looking to see if I'm going to hit myself with the yo-yo or the rope. So far, that hasn't happened.

I twirled the rope around my head and you saw that, of course, the rope became taut. And we interpret, typically, this as the centrifugal force. Something was pulling directly outward from the center of spin, which was my hand, pointing outward and holding it taut. And yet we're usually also taught that centrifugal force is false and only centripetal force is real. By now, we should be a little more comfortable dealing with forces that aren't strictly real, but we do need to ask, how can a force keeping the string taut not actually be real?

If you've been to an amusement park, you've often noticed that these amusement parks have this ride, a big cylinder. And you get inside the cylinder and once they close the door and trap you inside, the cylinder starts to rotate. And it rotates faster, and it rotates faster and faster, and as it does, you find yourself with your back stuck up against the wall. And then that's when the pain starts, and then as it's rotating very fast, they add insult to injury by dropping the floor away from you, showing that they are in total control. Now, I think I'm expressing my disdain and displeasure for this ride. I've done it exactly once, and to me, it felt like an invisible hand was pressed against my chest, pressing me against the wall, squeezing the air out of my lungs.

Let's say you're there instead, and you're feeling this invisible hand and it's pushing you. It's pressing you. And the pain is increasing and you finally, with your last breath in your lungs, you scream out, "I hate centrifugal force." But the person who's glued up against the wall next to you says, "Excuse me. I teach physics and I have to tell you that centrifugal force is not really real." You want to hit the guy, but you can't peel your arm off the wall. Is our physics professor right? Certainly, the pain is real, but is the source of the pain identified correctly?

Well, here you are in the cylinder ride, and we're looking down at your increasingly misshapen body. I'm showing the cylinder ride turning counterclockwise, but counterclockwise or clockwise, it doesn't matter. It's all the same pain. The key point to keep in mind is inertia. Wherever you are, inertia is always trying to make you move in a straight line, but the wall is preventing that. So in this case, you're actually literally constantly crashing into the wall. Now, we interpret this as a centrifugal force, acting outward like an unseen hand, directly outward from the center of spin, pushing us up against the wall.

But you know what? We know the centrifugal force is not real. We know it's not real because if suddenly, magically, the wall were to suddenly disappear, we know we would not fly directly outward in the direction of the centrifugal force. Instead, we'd fly off at a tangent, as inertia desires. So the important force here is provided by the wall, and the wall is pushing directly inward, and that's the centripetal force.

So here's a rule of thumb. Every time you infer a centrifugal force, it's really a centripetal force in disguise. It's like the difference between having a cup that's half-full of water. Is it half-full or half-empty? It's like the difference between partly cloudy and partly sunny. You know what? Even I don't know the answer to that one. But in my example with the rope, there was a centripetal force and it was produced by my hand and the string that I was holding, and actually, that was the force that was keeping the string taut.

In any event, we can interpret the third force in gradient wind balance around large-scale lows or highs as being centripetal acting inward, or centrifugal acting outward, whichever is easiest for us to visualize. For gradient wind balance, I'm going to choose the inward force because it's acting in the direction that the wind is trying to curve. So again, let's start with geostrophic balance, pressure gradient versus Coriolis force, and we're going to add in the centripetal force now, keeping in mind that the centripetal force always acts inward.

It's got to point either in the direction of pressure gradient or in the direction of Coriolis force. If I added in the direction of pressure gradient force, the combination of pressure gradient and centripetal forces joining forces against the Coriolis force is going to help guide the air counterclockwise around that low. Around the high, the inward-directed centripetal force has joined forces with Coriolis instead, guiding the parcel clockwise around the high. I could have done exactly the same explanation with outward-directed centrifugal. It's just a little less obvious why putting an extra force this way makes something curve that way.

So I've just given you the standard textbook explanation for gradient wind balance, leading to counterclockwise flow around Northern Hemisphere lows, clockwise flow around Northern Hemisphere highs. But I find this explanation unsatisfying because it causes a chicken and egg dilemma. Where did the centripetal and centrifugal forces come from? Air can't curve without a centripetal or a centrifugal force, but there is no such force before it starts to curve. So allow me to give you a different explanation, one that I find a lot more satisfying, one in which inertia plays a starring role.

Let's start again with geostrophic balance, the wind blowing with low pressure to the left. And again, I'll presume that the pressure gradient force is pointing north, but that's not strictly necessary. Suppose the air parcel's path is taking it towards a place where the isobars are curving in a counterclockwise fashion. Now, isobars can curve for many different reasons. We don't need to be concerned with that right now. The key point is inertia wants the air to continue moving straight, but notice this would carry the air across the 1000-millibar isobar towards higher pressure. Now, that's okay. The large-scale wind can flow across isobars. They're not material surfaces. They flow across them when friction is active.

But let's pretend the parcel is your car, and the isobars represent the roadway. And you want to stay on the roadway. You'd need to turn the wheel to stay in your lane. Well, the air parcel does have an automatic steering wheel of sorts, the pressure gradient force. If inertia had its way, and continued carrying the parcel straight, notice that the pressure gradient force, which always points most directly towards lowest pressure, is now not only pointing to the parcel's left, but also pointing to the parcel's back. If we zoom in on the parcel, we'll see that the pressure gradient force vector is 2 components, one part pointing to the left of the motion, just as before, but now the other component is pointing to its back. It's opposing the motion. Part of the wind's driving force is directed against the parcel.

Back to the car analogy; this is like slamming on the brake pedal. The air must slow down. So what happens is the air slows down. As the air slows down, Coriolis force is reduced. So here's another situation in which the Coriolis force is weakened, allowing the pressure gradient force to gain the upper hand, permitting it to change the direction. Except this time, it doesn't result in cross-isobar flow. Instead, the parcel remains following the isobars as they curve, keeping low pressure to the left, so we've made a cyclone with air curving counterclockwise around low pressure. But keep in mind one thing about this situation—it was necessary for the parcel to slow down, to curve to the left, because it required the Coriolis force to weaken relative to PGF.

Let's consider what happens if the parcel approaches isobars that are curving clockwise instead. It's a very similar situation. Again, inertia wants the parcel to continue straight, but this time, that would carry the parcel towards lower pressure, which is exactly what the parcel wants. The pressure gradient force vector is now turning, so a component of that vector is pointing in the direction of the parcel motion. This is you pressing the accelerator. The parcel is going to speed up. But this increases the Coriolis force, and Coriolis wants to make the air bend to the right. So it gains the upper hand on the pressure gradient force, and the air starts bending rightward. This keeps the air parallel to the isobars, but moving more quickly than it did when the motion was purely geostrophic. So to curve clockwise, it's necessary for the parcel to speed up.

In this explanation, constant curving of the flow clockwise or counterclockwise results from an imbalance between the pressure gradient and Coriolis forces, and we can interpret that imbalance as a centripetal or a centrifugal force, however we choose. However, did you notice that the air had to slow down to curve counterclockwise, and speed up to curve clockwise? I'll return to this asymmetry soon.

So now, let's combine our 4 fundamental forces. Pressure gradient force and Coriolis make straight line geostrophic flow, low pressure to the left. Centripetal gives us curvature, counterclockwise around low, clockwise around high, still parallel to isobars. Friction gives us a little cross-isobar flow into low and out of high. This is what the flow around large-scale lows and highs looks like near the surface. Notice that owing to friction, wind can cross isobars, away from high and towards low pressure, and note further, and this is important, note that it means that the air is converging into the low pressure from below. If I bring air into low pressure at the surface, it's easy to see that I will be creating upward vertical motion in the cyclone. Rising air can become saturated and possibly unstable, resulting in clouds and storms. And this is one of the reasons we associate cyclones with storms.

Meanwhile, the surface divergence out of the high implies downward motion in the anticyclone. And we usually associate highs with very stable and clear conditions. So we've looked at 4 fundamental forces influencing the wind so far, and we've actually examined 5 combinations thus far: pressure gradient force alone—that's what we did first, wind blowing directly from high pressure to low. But that only works when the length and timescales are relatively short. Pressure gradient and Coriolis, geostrophic balance, straight line parallel to isobars, isobar spacing indicating the magnitude of the pressure gradient force and the wind speed.

PGF, Coriolis, and friction—a modified geostrophic balance, allowing air to move at a shallow angle across isobars towards low pressure. PGF, Coriolis, and centripetal or centrifugal, the gradient wind balance, clockwise around highs, counterclockwise around lows. All 4 forces together—a modified gradient wind balance, with air spiraling counterclockwise into lows and spiraling clockwise out of highs. Remember, we said isobar spacing indicates wind speed because it reflects the magnitude of the pressure gradient force.

But we've also seen that isobar curvature affects wind speed as well, so that when, for the same isobar curvature, as the air curves counterclockwise, the wind is actually what we call "subgeostrophic." It is slower than if the isobars were straight and we had the geostrophic balance. In contrast, the flow clockwise around highs is supergeostrophic, and faster than we would have expected from the geostrophic balance. This is an apparent paradox because we often see strong winds around lows, and we associate highs with weak winds. In practice, isobar spacing around lows can be much smaller than around highs, and that's what leads to the faster winds.

We don't usually see tight pressure gradients around highs, and why is that? The reason is the sixth combination of these forces—pressure gradient and centrifugal force result in cyclostrophic balance, to turn in a circle, spin. Note Coriolis has no role in this. It's local-scale rapid spin. This is what describes the amusement park ride, as well as tornadoes and bathtub drains. And yet this illustrates a very powerful point—spin creates low pressure.

Let me demonstrate that with a beaker of water and a spoon. I'm starting off with an undisturbed fluid here, and if you think about the pressure at the bottom of the cup, which is our surface pressure, the surface pressure is the same everywhere. The same amount of mass of water and also atmosphere is pushing down, so there are no pressure gradient forces and there is no motion, but I can create a pressure gradient force. I can create a pressure gradient force pointing inward towards the center of the spin I'm about to introduce by stirring. I'm creating a vortex.

I'm creating a vortex. The spin has caused the fluid to pile up at the edges of the glass and dip down in the center. So now I have high pressure at the outside of the glass at the bottom here because the fluid is deeper, and low pressure in the center of the vortex. I have a pressure gradient force pointing in. And I can interpret this as a centrifugal force pointing out as well if that satisfies us. By the way, does it matter which way I stir the water? No. Do it for yourself. Counterclockwise and clockwise spin both create low pressure.

Actually, there's a clue there. With the large-scale wind, we only had counterclockwise flow around lows. We had clockwise flow around highs. But that's because the Earth's rotation was involved, the Coriolis force, and that's not true here. Instead, we're showing that spin by itself induces low pressure, and this has so many applications in meteorology. And the particular application at hand is it shows us why isobars can be so tightly packed around cyclones and why they are not tightly packed around anticyclones. Because of what I call the "cyclostrophic effect," spin supports low pressure. If I stir the fluid faster, the pressure gets lower. By the same token, the cyclostrophic effect if I stir winds clockwise around high pressure faster and faster and faster, what am I doing? I'm actually lowering the pressure in the center of that spin. I'm killing the anticyclone.

Let's summarize. In this lecture, we've seen the third and fourth fundamental forces that help determine how and how quickly the horizontal winds blow: friction and the centripetal or centrifugal force. Friction acts near the ground to reduce wind speeds. We saw that friction disturbs the geostrophic balance of pressure gradient and Coriolis forces, which is the wind that is not blowing towards low pressure, as Nature desires. It does it by acting directly and specifically against the Coriolis force, thereby letting the wind find its way towards low pressure, at least a little bit.

With the fourth force, we saw that we could choose to interpret it as centripetal or centrifugal, or centripetal versus something else, depending on what we found to be convenient. But the centripetal or centrifugal forces appear wherever there is spin. Centripetal forces act inward and are always real. Pressure gradient force is its usual disguise. Centrifugal forces act outward, and even though they're an illusion, they could be a convenient untruth, taking the blame when it's really centripetal force that's acting up. Gradient wind balance was made when this 2-headed force joined forces with pressure gradient and Coriolis forces. That explained why the wind turns counterclockwise around large-scale cyclones, areas of low pressure, and clockwise around anticyclones, or large-scale highs.

I also explained this in a different way, one that at least I liked a lot better. Adding friction to the gradient wind balance helped us understand one big reason why we associate low pressure systems with bad weather, or should I say good weather. Friction lets air converge into large-scale cyclones at the surface. This makes the air rise in the cyclone, push from below, and this air can diverge out of anticyclones, leading to descent, one reason why we associate highs with clear weather. We also addressed an interesting paradox. We saw that to make the large-scale wind deviate from the straight and narrow path and turn counterclockwise, we had to slow it down. We had to apply the brakes. And to make it turn clockwise, we had to speed it up and hit the gas pedal.

Circular flow around highs is supergeostrophic, and around lows, it's subgeostrophic. And yet we often see strong winds and tight gradients around lows, but almost never around highs. The answer came from cyclostrophic balance. However you want to formulate this force balance, this showed that spin makes low pressure. Cyclostrophic balance isn't just a tempest in a teacup, although it explained the tempest vortex. We're going to be seeing it everywhere because spin is ubiquitous.

Let's look ahead. In our next lecture, we will use our knowledge of how these fundamental forces combine to produce winds and wind directions to come to grips with a key question that we've been moving towards answering. What is the large-scale atmospheric circulation on a rotating Earth that has equator to pole temperature gradients? And as part of that, let's consider this: How did planetary rotation and sphericity combine to condemn the Sahara to be a barren wasteland? We'll see why in the next lecture.

Glossary

absolute vorticity: *See vorticity*.

adiabatic: A term referring to processes in which an object's temperature changes occur without heat exchange with the surrounding environment. See dry adiabatic and moist adiabatic. Greek: impassible.

advection: Spatial transport of some property by the winds, such as temperature (i.e., warm and cold advection) or vorticity. *See convection*.

air mass: A large body of air that forms over a particular region, acquiring certain characteristics; common air masses include continental polar (cP), continental tropical (cT), maritime polar (mP) and maritime tropical (mT) varieties.

aneroid barometer: A barometer that uses elastic membranes to measure pressure. Greek: without fluid.

anticyclone: A large-scale region of high pressure characterized by clockwise (CW) flow in the Northern Hemisphere. Latin: opposite of cyclone. Coined by Sir Francis Galton.

barometer: An instrument used to measure pressure. Greek: to measure weight.

beta drift: Mechanism of self-propagation of a tropical cyclone owing to Earth's curvature.

blackbodies: Objects that absorb all incident radiation; blackbodies do not have to be colored black, but that helps.

Bulk Richardson number (BRi or BRN): The ratio between CAPE and vertical wind shear, useful in severe weather forecasting.

Buys Ballot's law: "In the northern hemisphere, with the wind at your back, lower pressure will be to the left." An empirical description of the geostrophic and gradient winds; named for C. H. D. Buys Ballot, a Dutch meteorologist.

centrifugal force: An apparent force directed outward from the center of spin. Latin: to flee the center.

centripetal force: A spin-related force acting towards the center of spin. Latin: to seek the center.

chlorofluorocarbons (CFCs): A family of greenhouse gases that were also implicated in the enlargement of the ozone hole; CFCs are or were used as propellants in spray cans, coolants in refrigerators and air conditioners, solvents and fire extinguishers.

conduction: Heat transfer by direct molecular contact; involves transferring microscopic KE, measured by T, from the object with more to the object with less.

convection: Generally, heat transport by mass fluid motion. "Free convection" is self-driven, by positive buoyancy; "forced convection" requires an external agent. "Convection" is used as a synonym for atmospheric circulations driven by latent heat exchange. "Convective initiation" refers to the genesis of thunderstorms, which are composed of "convective cells." Latin: to carry, to convey.

convective available potential energy (CAPE): A measure of positive buoyancy potentially available to a rising air parcel between its level of free convection and equilibrium levels; fuel for storms.

convective inhibition (CIN): A measure of negative buoyancy encountered by a rising air parcel below its level of free convection.

Coriolis force: Apparent force owing to Earth rotation, acting to the right following the motion in the northern hemisphere; named for French scientist Gaspard-Gustave Coriolis.

crepuscular rays: An optical phenomenon in which light is partially blocked by opaque objects, such as clouds, trees, etc. Latin: relating to twilight.

cyclone: A region of low pressure with winds traveling in a closed, curved path. For large-scale cyclones in the northern hemisphere, this implies counterclockwise flow representing gradient wind balance, in contrast to anticyclones. Types include extratropical cyclones and tropical cyclones, including hurricanes and typhoons. The flow around small-scale cyclones, such as mesocyclones, may be counterclockwise or clockwise and represents cyclostrophic balance. Greek: circle, or circular path.

cyclostrophic balance: The combination of PGF and centrifugal forces; Greek: to turn in a circle.

density: Mass over volume; measured in kilograms per cubic meter.

derecho: A type of squall line in which the thunderstorm band becomes bowed in shape (i.e., bow echoes), often associated with intense, damaging straight-line winds. Spanish: straight.

dew point temperature (Td): The temperature at which air saturates by diabatic cooling. Along with the pressure, Td tells us the air's vapor supply.

diabatic: Temperature change caused by heat exchange—the addition or extraction of heat energy. Greek: passible.

diffluence: Horizontal spreading of an airflow at a particular level with distance and time; can be confused with divergence. Latin: to flow apart.

divergence: Horizontal spreading of an airflow at a particular level that is not compensated by an equal amount of slowing, resulting in ascending motion beneath the level and/or descending motion above.

dry (subsaturated) adiabatic: The process in which temperature changes solely due to volume and pressure changes, resulting in expansion cooling or compression warming.

dry adiabatic lapse rate (DALR): The rate in which temperature decreases (lapses) with height due to the dry adiabatic process; fixed at 10° C per kilometer or roughly 30° F per mile.

dry line: A moisture boundary common in the western part of the American Central Plains during spring and summer seasons, often involved in convective initiation.

Ekman spiral: Refers to the turning of the ocean current direction with depth in the ocean surface layer. Ekman transport refers to the net motion of the ocean current over the depth of the Ekman spiral. Discovered by Swedish oceanographer V. W. Ekman.

El Niño: A weather and ocean pattern during which the eastern tropical Pacific is warmer than normal. Spanish: the boy child. Cooler than normal periods are dubbed La Niñas. Spanish: the girl child.

electromagnetic spectrum: Comprises the range of electromagnetic radiation, differentiated by wavelength λ. From small wavelength to large, regions of the spectrum include gamma and X-rays ($\lambda < 10-9$ m); UV rays ($\lambda \sim 10-7$ m); visible light (0.4–0.7 μm); infrared; microwaves and radio waves.

environmental lapse rate (ELR): The rate at which temperature decreases (lapses) with height; varies with space and time, but averages 6.5° C per kilometer or 20° F per mile in the standard atmosphere's troposphere. When temperature increases with height, the ELR is negative.

equilibrium level (EQL): The level at which the positive buoyancy of a rising air parcel vanishes; also called a cloud top.

front: A boundary between air masses having different densities. Large-scale fronts include cold, warm, stationary, and occluded fronts. Other types include sea breeze and gust fronts.

Froude number: The ratio between horizontal velocity and the shallow water wave speed, useful in diagnosing downslope wind situations $Fr = \dfrac{u}{\sqrt{gD}}$.

Fujita scale: A scale devised by Theodore Fujita to classify the intensity of tornadoes, based on resulting damage consiting of 6 categories: F0 to F5. Recently succeeded by the Enhanced Fujita (EF) scale.

gradient wind balance: The balance among pressure gradient, Coriolis, and centripetal forces, resulting in counterclockwise (clockwise) flow around northern hemisphere cyclones (anticyclones).

geopotential height: The height of a pressure level above mean sea level; in gravity-adjusted units that, in practice, is very close to geometric meters. Unit is geopotential meters (gpm).

geostrophic balance: The balance between pressure gradient and Coriolis forces resulting in the wind blowing in a straight line with low pressure to the left in the northern hemoisphere. Greek: because the Earth turns.

greenhouse effect: Refers to the fact that certain constituents of the Earth's atmosphere—the greenhouse gases, including water vapor, carbon dioxide CO_2, methane CH_4, nitrous oxide N_2O, and ozone O_3—selectively absorb and reradiate longwave radiation, resulting in the Earth's surface and atmosphere being warmer than they otherwise would have been.

Hadley cell: The tropical vertical circulation cell characterized by ascent at the equator and descent at 30° latitude, part of the 3-cell model.

heat: The flow of energy between objects.

heat conductivity: A measure of a substance's ability to conduct heat. Metals have high heat conductivity; air is a very poor conductor.

hydraulic jump: A fluid phenomenon characterized by a sharp change in depth and flow speed, often very turbulent.

hydrostatic balance: The stalemate between the vertical pressure and gravity forces resulting in no (accelerated) motion. Greek: balanced fluid.

ideal gas law: $p = \rho rt$; p is pressure, measured in pascals; ρ is density; t is temperature in the Kelvin scale; and r is a proportionality constant unique to each gas or combination of gases.

infrared (IR): A portion of the electromagnetic spectrum between the visible and microwave regions, divided into 2 sections: near IR (0.7–1.5 μm) and far IR (> 1.5 μm). Latin under red.

inter-tropical convergence zone (ITCZ): Meeting place betweenn and southern hemispheric air.

iso-: from Greek, a prefix used to indicate equal, used in isobars (lines of equal pressure), isotherms (temperature), isotachs (speed), and isoheights. Variant form used in isentropes (entropy).

katabatic: Referring to a wind blowing downslope. Greek: to flow downhill.

Keeling curve: A plot of atmospheric CO_2 concentration with time, named after Charles David Keeling, the scientist who started routine measurement of this gas.

knot (kt): A nautical mile per hour. 1 kt = 1.15 mph = 0.51 m/s = 1.85 km/h.

latent heat: Refers to heat absorbed or released by a substance involved in phase transitions. In evaporation, latent heat is the energy used to break molecular bonds, liberating water molecules. That heat is returned to the water substance's surroundings when the bonds reform during condensation. Latin: to be hidden.

©2010 The Teaching Company.

level of free convection (LFC): The level at which a rising air parcel is slightly warmer than its surrounding environment, and thus becomes positively buoyant.

lifting condensation level (LCL): The level at which air can be brought to saturation by lifting, the dry adiabatic approach to saturation; cloud base.

longwave radiation: Radiation emitted by typical Earth objects, being wavelengths > 3 μm and thus consisting of far IR, microwaves, and beyond.

mesoscale convective system (MCS): A large, organized thunderstorm complex, with length and time scales much larger than those associated with individual thunderstorm cells. Examples: squall lines and supercell storms.

mesosphere: Atmospheric layer above the stratopause, characterized by temperature again decreasing with height. Top is called the mesopause. Greek "middle sphere."

meteorology: Greek for "the study of things high in the sky"; the title of a book written by Aristotle around 350 B.C.

millibar (mb): The traditional unit of pressure in meteorology, representing one thousandths of a bar, introduced by British meteorologist Sir William Napier Shaw and formally succeeded by the hectopascal (hPa). Standard sea-level pressure is 1013.25 mb or 1013.25 hPa. *See pressure.*

moist (saturated) adiabatic: The process in which an air sample's temperature changes due to expansion or compression modified by water substance phase changes, but still without heat exchange with the surroundings.

moist adiabatic lapse rate (MALR): The rate at which the temperature of a saturated air sample decreases (lapses) with height owing to the moist adiabatic process. The MALR is extremely variable, ranging from 3° C/km for hot air to almost 10° C/km for very cold saturated air. A reasonable representative average is 5° C/km or 15° F/mi.

monsoon: The seasonal reversal of winds, which often result in rainfall variations and often associated with weather patterns in India and Southeast Asia. Arabic: season.

Newton's second law: Force equals mass times acceleration ($F=ma$). In the metric system, force is measured in newtons (N), which is the mass unit (kilogram) multiplied by acceleration (meter per second squared).

Newton's law of inertia: An object, once placed in motion, remains moving in a straight line with constant speed unless other forces are acting.

nucleation: Aggregation of water molecules during phase changes. The former requires condensation nuclei, especially hygroscopic particles such as dust, sand, salt, and soot. For temperatures above $-40°$ C = $-40°$ F, freezing requires an ice nucleus (heterogenous nucleation), which can be scarce. Cloud seeding involves artificially introducing ice nuclei.

1-cell model: A circulation in a vertical plane in which one branch of ascending air and one branch of descending air combine to form a closed circulation cell.

ozone (O_3): A rare trace gas in the atmosphere, composed of 3 oxygen atoms and located mainly in the stratosphere where it absorbs harmful radiation that would otherwise reach the ground. Greek: to smell.

ozone hole: Refers to a region of enhanced seasonal ozone depletion over the South Pole exacerbated chemical reactions involving man-made compounds, notably chlorofluorocarbons (CFCs).

Planck's law: Tells us how much of each kind of radiation an object produces. The graph of radiative energy emitted versus wavelength is called the Planck curve.

planetary waves: Large-scale variations in pressure on constant-height charts, or height on isobaric charts. Axes of lower (higher) height or pressure are termed troughs (ridges). Also called Rossby waves.

positive vorticity advection (PVA): Horizontal transport of positive absolute vorticity by the winds.

pressure gradient force (PGF): Largely determine wind speeds; a p gradient = p difference over distance; in addition to other forces: friction, Coriolis, and centripetal.

pressure: Force per unit area; in the atmosphere, pressure is largely produced by the weight of overlying air pressing downward owing to gravity. Units include the pascal (1 Pa = 1 Newton per square meter) and hectopascal (100 Pa), bar and millibar, pound per square inch, and inch of mercury.

radar: A system that uses electromagnetic waves to detect the range and quantity of objects capable of backscattering the wavelength employed. Weather radars tend to use wavelengths of 10 centimeters for long-range precipitation radar, and ~3 centimeters for cloud radar. Originally, an acronym for RAdio Detection And Ranging.

radiation: A means of heat transfer in which energy travels as waves at the speed of light. *See electromagnetic spectrum.*

Rayleigh scattering: Refers to the theory of how electromagnetic radiation behaves upon encountering particles smaller than the wavelength of said radiation, used to explain the blue sky and reddish setting Sun, and postulated by English physicist Lord Rayleigh.

relative humidity: The ratio between vapor supply and vapor capacity.

refraction: Refers to a change of direction of a wave, such as light, owing to speed change.

Saffir-Simpson intensity scale: A 5-category classification system used for Atlantic and eastern Pacific hurricanes developed by Herbert Saffir and Robert Simpson, based on maximum sustained (1 min) wind speed measured at 10 m (33 ft.) above the ground.

Santa Ana winds: A warm, dry katabatic wind common in southern California in the fall and winter seasons, in which originally cold, dense air warms dry adiabatically through descent, decreasing the relative humidity in the process.

shortwave (solar) radiation: Radiation produced by hot objects such as the sun, being wavelenghs less than 3 μm and thus consisting of ultraviolet and shorter rays, visible light, the near IR, and part of the far IR.

squall lines: Narrow linear or curvilinear bands of thunderstorms typically characterized by heavy precipitation and strong winds and often trailed by an extensive region of lighter "stratiform" precipitation.

Stefan-Boltzmann law: The relationship between temperature T and radiative energy output E. E is proportional to the fourth power of T.

stratosphere: The layer of the atmosphere overlying the troposphere, in which temperature either increases with height or ceases decreasing, owing to absorption of ultraviolet radiation by oxygen and ozone. Top is called the stratopause. Latin "to spread (horizontally)."

supercell: A type of thunderstorm characterized by horizontally rotating updrafts, often associated with tornadoes.

supercooling: The process of lowering the temperature of liquid water below the nominal freezing point. Supercooled liquid can persist when there is a shortage of ice nuclei and is involved in aircraft icing.

temperature (T): A measure of the microscopic kinetic energy of atoms and molecules, which vibrate and translate, even in solids, so long as T greater than absolute zero.

temperature inversion: A situation in which temperature increases with height.

thermal inertia: A measure of a substance's resistance to temperature change. Metals have low thermal inertia; liquid water's inertia is large. Also termed heat capacity.

thermally direct circulation: Circulations driven by horizontal temperature differences in which adiabatic vertical motions contribute to decreasing the temperature contrast, such as the sea breeze in which the ascending warmer air cools via expansion and the sinking cooler air warms by compression. The vertical motions in thermally indirect circulations enhance the temperature contrast.

thermosphere: Upper portion of the Earth's atmosphere, from the mesopause to where atmosphere fades away.

3-cell model: A simple conceptual model of the atmospheric circulation in a vertical plane stretching from equator to pole involving ascent at the equator and 60° latitude, and descent at the pole and 30° latitude. The 3 cells are termed the Hadley, Ferrel, and Polar cells.

troposphere: The lower part of the Earth's atmosphere, in which temperature generally decreases with height. Greek: "turning sphere."

vapor supply (VS) and **vapor capacity (VC)**: Quantify the present and saturation values of water vapor, expressed as mixing ratios; measured in grams of water per kilogram of dry air.

vorticity: The condition of spinning or vortical motion, identified by the orientation of the spin axis. Vertical vorticity is spin in a horizontal plane, positive for counterclockwise motion. Planetary vorticity (f) is the component of the Earth's spin in the local vertical. Relative (vertical) vorticity (ζ) is horizontal spin relative to the Earth. Absolute or total vorticity is f + ζ, which is conserved in the absence of sources and sinks. Latin: whirl.

Walker circulation: An east-west circulation system in the tropics, deduced by Sir Gilbert Walker.

wet bulb temperature (Tw): The temperature at which air can be saturated via evaporation of liquid water.

Wien's law: The wavelength of maximum radiative emission for an object varies inversely with its temperature, identifying the peak of the Planck curve.

wind chill: Accelerated heat loss owing to convection.

wind shear: The variation of the speed and/or direction of the horizontal wind over distance, either horizontally or vertically.

Z (as in **18Z Oct 31**): A commonly used abbreviation for Greenwich Mean Time (GMT), also known as Coordinated Universal Time (UTC), which is based on standard time at the Greenwich Observatory in London, UK.

Bibliography

This course was pitched at the freshman/sophomore level, with a few more advanced topics incorporated here and there. A good place to begin further reading is with the books listed in the introductory-level texts section. The popular books section includes a few accessible treatments of specific subjects covered in this course. The advanced texts are used in undergraduate and/or graduate-level courses that have calculus and physics prerequisites. Some of the scientific literature articles are very technical, others less so. Those articles mainly focus on topics covered in the latter third of the course.

Introductory-Level Texts:

There are many high quality introductory meteorology texts suitable for general education meteorology courses. Ahrens's *Meteorology Today* (along with its streamlined cousin, *Essentials of Meteorology*) is probably the best known; I've used them in freshman classes. Lutgens et al., Aguado and Burt, and Ackerman and Knox cover much the same territory. Danielson et al. appears to be out of print, but is distinguished by a much more "weather"-focused presentation. Rauber et al. focus more narrowly on severe weather in all seasons and does it very well. Successive editions of these texts are generally little changed so older versions are both good and cheap!

Ackerman, Steven A., and John A. Knox. *Meteorology: Understanding the Atmosphere*. 2nd edition. Brooks Cole, 2007.

Aguado, Edward, and James E. Burt. *Understanding Weather and Climate*. 5th edition. New York: Prentice Hall, 2009.

Ahrens, C. Donald. *Meteorology Today: An Introduction to Weather, Climate, and the Environment*. 9th edition. Brooks Cole, 2008.

Ahrens, C. Donald. *Essentials of Meteorology*. 5th edition. Brooks Cole, 2008.

Danielson, Eric W., James Levin, and Elliot Abrams: *Meteorology*. 2nd edition. New York: McGraw-Hill Education, 2002.

Lutgens, F. K,, E. J, Tarbuck, and D. Tasa. *The Atmosphere: An Introduction to Meteorology*. 11th edition. New York: Prentice Hall, 2009.

Rauber, Robert M., John E. Walsh, and Donna Jean Charlevoix. *Severe and Hazardous Weather*. 3rd edition. Dubuque, IA : Kendall/Hunt, 2008.

Popular Books on Specific Subjects:

Comins, Neil F. *What If the Moon Didn't Exist? Voyages to Earths That Might Have Been.* New York: Harper Collins, 1994. Although out of print, this book presents a wonderful set of thought experiments, including the title exercise in which he discusses the role of the Moon in shaping Earth's weather and the evolution of life.

Emanuel, Kerry. *Divine Wind: The History and Science of Hurricanes.* New York: Oxford, 2006. What we know (and don't know) about hurricanes and typhoons, presented in a very accessible manner (Lecture Twenty-One).

Gleick, James. *Chaos: Making a New Science.* New York: Penguin, 1988. A very entertaining take on Ed Lorenz and the science of "chaos theory" he helped spawn.

Hoeppe, Götz. *Why the Sky Is Blue: Discovering the Color of Life.* New York: Princeton University Press, 2007. Hoeppe discusses the role of ozone in the color of the twilight sky, among other appreciations of the beautiful color blue.

Lynch, Peter. *The Emergence of Numerical Weather Prediction: Richardson's Dream.* Cambridge: Cambridge University Press, 2006. An accessible account of the early days of numerical weather forecasting.

Stewart, George Rippey. *Storm.* Berkeley, CA: Heyday Books, 2003. The novel, famous in its own right, but also generally credited with inspiring the practice of assigning names to storms.

Turner, Gerard L'E. *Scientific Instruments 1500–1900: An Introduction.* Berkeley: University of California Press, 1998. Turner recounts the story of the Fahrenheit and Celsius thermometers, and discusses many other scientific instruments.

Vonnegut, Kurt. *Cat's Cradle.* New York: Penguin, 1998. Kurt, younger brother of cloud-seeing pioneer Bernard Vonnegut, contemplated what would happen if an ice nucleus active above the freezing point were to escape from the laboratory.

Advanced Texts:

Bohren, Craig F., and Bruce A. Albrecht. *Atmospheric Thermodynamics.* New York: Oxford University Press, 1998. All about thermodynamics, especially involving water substance. The discussion of why we are not crushed by overflying aircraft is on page 73.

Holton, James R. *An Introduction to Dynamic Meteorology.* 4th edition. Burlington, MA: Elsevier, 2004. The most famous dynamic meteorology textbook; not for the timid.

Houze, Robert A., Jr. *Cloud Dynamics.* 1st edition. San Diego, CA: Academic Press, 1993. Rigorous but readable treatment of clouds and related phenomena, including supercell storms and tornadoes (chap. 8), squall lines (chap. 9), hurricanes (chap. 10), and clouds associated with fronts and mountains (chaps. 11–12).

Kocin, Paul J., and Louis W. Uccellini. "Northeast Snowstorms. Volume 1: Overview," *Meteorological Monographs,* vol. 32. no. 4. (2004). All about snowstorms of the Northeastern United States, relevant to Lectures Fourteen to Sixteen.

Martin, Jonathan E. *Mid-Latitude Atmospheric Dynamics: A First Course.* New York: John Wiley and Sons, 2006. This text provides a rigorous but well-explained treatment of the extra-tropical cyclone, including concepts found in Lectures Fifteen and Sixteen.

Stull, Roland B. *Meteorology for Scientists and Engineers.* 2nd edition. San Diego, CA: Brooks/Cole, 1999. Originally designed as a technical companion for Ahrens's *Meteorology Today,* this text provides the physical and mathematical basis for fundamental meteorological phenomena.

Wallace, John Michael, and Peter Victor Hobbs. *Atmospheric Science: An Introductory Survey.* 2nd edition. Academic Press, 2006. A revised version of the classic textbook, and a good place to start for a more advanced treatment of meteorology.

Scientific Literature:

A few selected papers, mainly concentrating on storms, along with some background for the hurricane track experiment discussed in Lecture Twenty-Four. Technical content varies.

Bluestein, H. B., and M. H. Jain. "Formation of Mesoscale Lines of Precipitation: Severe Squall Lines in Oklahoma during the Spring," *Journal of the Atmospheric Sciences* 42:1711–1732, (1985). This classic paper helps distinguish between supercell storms and various kinds of squall-lines.

Browning, K. A., et al. "Structure of an Evolving Hailstorm, Part V: Synthesis and Implications for Hail Growth and Hail Suppression," *Monthly Weather Review,* 104:603–610 (1976). A famous depiction of the structure of the multicellular thunderstorm.

Byers, H. R., and R. R. Braham, Jr. "Thunderstorm Structure and Circulation." *Journal of the Atmospheric Sciences*, v. 5, pp. 71–86, 1948. The most famous article to emerge from the Thunderstorm Project, depicting the life cycle of the thunderstorm cell.

Durran, D. R. "Mountain Waves and Downslope Winds," in *Atmospheric Processes Over Complex Terrain. Meteorological Monographs*, vol. 23, no. 45 (1990). All about mountain waves, including lee waves, hydraulic jumps, lenticular clouds, etc., especially pertinent to Lecture Seventeen.

Fovell, R. G., K. L. Corbosiero, and H. -C. Kuo. "Cloud Microphysics Impact on Hurricane Track as Revealed in Idealized Experiments," *Journal of the Atmospheric Sciences*, 66:176–1778 (2009). Why cloud processes can influence the direction and speed of tropical cyclones, relevant to Lecture Twenty-Four.

Fovell, R. G., and Y. Ogura. "Numerical Simulation of a Midlatitude Squall Line in Two Dimensions," *Journal of the Atmospheric Sciences*, 45:3846–3879 (1988). Squall line thunderstorms as self-organizing and self-perpetuating.

Fovell, R. G., and P.-H. Tan. "The Temporal Behavior of Numerically Simulated Multicell-type Storms. Part II: The Convective Cell Life Cycle and Cell Regeneration," *Monthly Weather Review*, 126: 551–577 (1998). Describes why multicell storms are unsteady.

Houze, R. A., Jr. "Mesoscale Convective Systems," *Reviews of Geophysics*, vol. 42 (2004). A readable survey of our understanding of MCSs as of 2004.

Lemon, L. R., and C. A. Doswell III. "Severe Thunderstorm Evolution and Mesocyclone Structure as Related to Tornadogenesis," *Monthly Weather Review*, 107:1184–1197 (1979). A famous description of the airflow around supercellular thunderstorms.

Leopold, L. B, "The Interaction of Trade Wind and sea Breeze, Hawaii," *Journal of Meteorology*, 6:312–320 (1949). A famous early paper on the sea breeze circulation.

Lorenz, E. A., "Deterministic Nonperiodic Flow," *Journal of the Atmospheric Sciences*, 20:130–141 (1963). This landmark paper that gave birth to chaos theory is not easily readable.

Rotunno, R., and J. B. Klemp, "The Influence of the Shear-Induced Pressure Gradient on Thunderstorm Motion," *Monthly Weather Review*, 110:136–151 (1982) Why supercell storm split and move at an angle to the mean wind.

Rotunno, R., J. B. Klemp, and M. L. Weisman, "A Theory for Strong, Long-Lived Squall Lines," *Journal of the Atmospheric Sciences*, 45:463–485 (1988). Why squall line storm updrafts tilt against the vertical shear vector, as discussed in Lecture Eighteen.

Weisman, M. L., and J. B. Klemp, "The Dependence of Numerically Simulated Convective Storms on Vertical Wind Shear and Buoyancy," *Monthly Weather Review*, 110:504–520 (1982). Environmental conditions favoring rotating supercell or multicellular storms.

Websites:

Archived Data and Tools:

Archived radar data: http://www4.ncdc.noaa.gov/cgi-win/wwcgi.dll? WWNEXRAD~Images2

Archived satellite data: http://www.class.ngdc.noaa.gov/saa/ products/welcome

Archived surface data: http://www7.ncdc.noaa.gov/CDO/cdo

Climate maps: http://cdo.ncdc.noaa.gov/cgi-bin/climaps/climaps.pl

Composite map creation tool (I used this frequently in the course): http://www.cdc.noaa.gov/cgi-bin/data/composites/printpage.pl

The NOAA photo library (hundreds of classic photographs): http://www.photolib.noaa.gov/

Past weather maps from the National Oceanic and Atmospheric Administration (NOAA):

http://www.hpc.ncep.noaa.gov/dailywxmap/index.html

http://docs.lib.noaa.gov/rescue/dwm/data_rescue_daily_weather_maps .html

Current Weather:

The National Weather Service: http://weather.gov/

The National Weather Service Doppler Radar page: http://radar.weather.gov/

The NWS Storm Prediction Center: http://www.spc.noaa.gov/

Current and Archived Surface Station Reports and Maps:

Oklahoma mesonet: http://www.mesonet.org/

Unisys weather: http://weather.unisys.com/

Good Place to Find Upper Air Charts and Model Forecasts:

Real-time weather from the National Center for Atmospheric Research: http://www.ral.ucar.edu/weather/

Hurricanes, Typhoons, and Tropical Meteorology:

Central Pacific Hurricane Center: http://www.prh.noaa.gov/hnl/cphc/

The Joint Typhoon Warning Center: http://metocph.nmci.navy.mil/jtwc.php

The National Hurricane Center: http://www.nhc.noaa.gov/

The U.S. Navy tropical cyclone page: http://www.nrlmry.navy.mil/tc_pages/tc_home.html

Satellite Pictures, Radar Images, Surface and Upper Air Maps Forecasts:

College of DuPage: http://weather.cod.edu

Plymouth State College Weather Center: http://vortex.plymouth.edu/

University of Illinois WW2010 site: http://ww2010.atmos.uiuc.edu/(Gh)/home.rxml

The University of Wyoming (atmospheric soundings in real time and archived): http://weather.uwyo.edu/upperair/sounding.html

Severe Weather Forecasts and Storm Damage Reports:

NWS Hydrometeorological Prediction Center (HPC): http://www.hpc.ncep.noaa.gov/

Surface Maps and Precipitation Forecasts:

HPC's current weather map: http://www.hpc.ncep.noaa.gov/dailywxmap/

MesoWest at the University of Utah: http://mesowest.utah.edu/index.html

Notes